An introduction to linear electric circuits

An introduction to linear electric circuits

L. V. Kite, MSc, PhD

Lecturer in Physics
Chelsea College, University of London

Longman

Longman
1724 - 1974

Longman Group Limited
Burnt Mill
Harlow
Essex CM20 2JE

Distributed in the United States of America by Longman Inc., New York

*Associated companies, branches and representatives
throughout the world*

© Longman Group Limited, 1974

First published 1974

ISBN 0 582 44256 7
Library of Congress Catalog Card Number: 73–85682

Printed in Great Britain by
Whitstable Litho Ltd

Contents

PREFACE   xii

ACKNOWLEDGEMENTS   xiv

1	Historical background	1
1.1	Introduction	1
1.2	Static electricity	1
1.3	Magnetism	2
1.4	Capacitance and charge	2
1.5	Electrochemistry	3
1.6	Electromagnetics	4
1.7	Electrical measurements	5
1.8	Telegraphy	6
1.9	Electromagnetic induction	6
1.10	Electric lighting and mains supplies	7
1.11	Electrical units	8
	Additional Reading	8

2	Basic electrical quantities	9
2.1	Electric current	9
	2.1.1 Current flow indicators	9
	2.1.2 Establishment of scale of current	10
	2.1.3 Kirchhoff's first law	12
	2.1.4 Electric charge	12
	2.1.5 Conservation of charge	13
2.2	Establishment of scale of voltage	13
2.3	Resistance	15
	2.3.1 Ohm's law	15
	2.3.2 Establishment of scale of resistance	16

2.4   Electromotive force   16
2.5   Kirchhoff's second law   17
2.6   Energy and power   18
       2.6.1   Conservation of energy   19
2.7   Electrical units   19
       2.7.1   Unit of current   20
       2.7.2   Unit of potential difference   21
       2.7.3   Unit of resistance   21
       References   22
       Examples   22

3   **Steady current circuit analysis**   23

3.1   Resistance   23
       3.1.1   Resistances in parallel   23
       3.1.2   Resistances in series   24
       3.1.3   Resistivity and conductivity   25
3.2   Symbols and sign conventions   26
3.3   Mesh currents   28
3.4   Sign conventions and Kirchhoff's second law   28
3.5   Worked examples   30
3.6   Maximum power theorem   32
3.7   Ladder networks   34
       3.7.1   Matched ladder   34
       3.7.2   Ladder attenuator   36
       3.7.3   Defects and limitations   37
       Examples   38

4   **Steady current measurements**   43

4.1   Standard cell   43
4.2   Resistors   44
       4.2.1   Standard resistors   44
       4.2.2   Precision commercial resistors   45
       4.2.3   Decade resistance boxes   46
       4.2.4   Radio-type resistors   46
       4.2.5   Variable resistors and potential dividers   47
4.3   Instrument shunts and series resistors   48
       4.3.1   Voltmeters   48
       4.3.2   Ammeters   49
       4.3.3   Universal shunt   50
       4.3.4   Multi-test instruments   51
4.4   DC potentiometers   51

		4.4.1	Simple potentiometer	52
		4.4.2	Crompton potentiometer	54
		4.4.3	Kelvin—Varley potential divider	55
		4.4.4	A modern potentiometer	55
		4.4.5	Potentiometer measurements	58
4.5	Wheatstone bridge			59
		4.5.1	Balance condition	60
		4.5.2	Balancing procedure	61
		4.5.3	Sensitivity	61
4.6	Carey—Foster bridge			62
4.7	Kelvin double bridge			63
	Additional reading			64
	Examples			65

| **5** | **Transients** | | | **68** |

5.1	Introduction			68
5.2	Capacitance			69
		5.2.1	Stored energy	70
		5.2.2	Capacitors in parallel	72
		5.2.3	Capacitors in series	73
		5.2.4	Discharging through resistance	75
		5.2.5	Charging through resistance	77
		5.2.6	Multi-loop networks	79
		5.2.7	Practical capacitors	82
5.3	Inductance			83
		5.3.1	Transformer effect	84
		5.3.2	Self induction	85
		5.3.3	Coefficient of self inductance	86
		5.3.4	Stored energy	87
		5.3.5	Inductors in series	88
		5.3.6	Inductors in parallel	89
		5.3.7	Growth in series RL circuit	90
		5.3.8	Decay in series RL circuit	92
		5.3.9	Sign convention in transformer effect	93
		5.3.10	Coefficient of mutual inductance	94
		5.3.11	Mutually coupled series coils	96
		5.3.12	A worked example	97
5.4	Series LCR circuits			99
		5.4.1	Equation of charge flow	99
		5.4.2	Transient initiation	100
		5.4.3	Overdamped circuit	101

	5.4.4	Underdamped circuit	103
	5.4.5	Critically damped circuit	106
	5.4.6	Charge decay	108
	5.4.7	Current decay	110
	5.4.8	Charge growth	111
5.5	Calibration of ballistic galvanometer		113
5.6	Integration and differentiation		115
	Examples		117

6   Alternating currents   122

6.1	Introduction		122
6.2	Resistive networks		124
	6.2.1	Current–voltage relation	124
	6.2.2	Power and energy	125
	6.2.3	Root-mean-square quantities	126
	6.2.4	Series and branched circuits	127
6.3	AC instruments		128
6.4	Capacitance		130
6.5	Inductance		132
6.6	Linearity		134
6.7	Power factor		134
6.8	Basic ac circuit analysis		135
	6.8.1	The complex exponent	135
	6.8.2	Resistance	136
	6.8.3	Capacitance	137
	6.8.4	Complex impedance	137
	6.8.5	Inductance	139
6.9	Series LCR circuit		140
	6.9.1	Resistance and reactance	141
	6.9.2	Argand diagram	141
6.10	Parallel LCR circuit		142
	6.10.1	Complex admittance	143
	6.10.2	Conductance and susceptance	144
6.11	Imperfect capacitor		144
6.12	Power matching		146
6.13	Worked examples		147
	Examples		153

7   Filters and transmission lines   157

| 7.1 | Filters | | 157 |
| | 7.1.1 | Matching load | 157 |

7.1.2	Propagation constant for matched ladder	158
7.1.3	Matched lossless ladder	159
7.1.4	Filters	161
7.2	Transmission lines	162
7.2.1	Basic considerations	163
7.2.2	Representative line element	163
7.2.3	Characteristic impedance	164
7.2.4	Equations of telegraphy	165
7.2.5	Phase-velocity	166
7.2.6	Line constants	168
7.2.7	Current and voltage distributions	170
7.2.8	Input impedance	171
	Examples	172

8	**Resonance**	**174**
8.1	General considerations	174
8.2	Series resonance	174
8.2.1	Resonant frequency	175
8.2.2	Voltage gain	176
8.2.3	Half-power points	177
8.2.4	Quality factor and selectivity	179
8.3	Q-meter	180
8.3.1	Susceptance measurement	181
8.3.2	Conductance measurement	182
8.4	Quality factor for nonresonant elements	183
8.5	Parallel resonance	184
8.5.1	Parallel combination with shunting resistance	185
8.5.2	Parallel combination with series resistances	189
	Examples	192

9	**Transformers**	**194**
9.1	General considerations	194
9.1.1	Voltage ratio	195
9.1.2	Coupling factor	195
9.1.3	Loss mechanisms	196
9.1.4	Open-circuit secondary voltage	196
9.1.5	Basic equations	197
9.1.6	Input impedance	198
9.1.7	Resistive loading	199
9.1.8	Equivalent primary circuit	200
9.1.9	Worked example	201

9.2	Tuned transformer	201
	9.2.1 Primary tuning	202
	9.2.2 Input impedance of tuned primary	204
	9.2.3 Tuning of secondary	204
	9.2.4 Symmetrical tuned transformer	206
	9.2.5 The principal solution	207
	9.2.6 The large coupling solutions	208
	9.2.7 Response curves	209
	9.2.8 Practical considerations	211
	Additional reading	211
	Examples	211

10	**Alternating current bridges**	**213**
10.1	The bridge principle	213
10.2	Wheatstone-type bridges	214
	10.2.1 Balance conditions	214
	10.2.2 Owen bridge	215
	10.2.3 Schering bridge	218
	10.2.4 Robinson frequency bridge	219
	10.2.5 Other Wheatstone-type bridges	221
	10.2.6 Layout	224
	10.2.7 Screening	224
	10.2.8 Wagner earth	227
	10.2.9 Screened and balanced transformer	229
	10.2.10 Limitations of Wheatstone-type bridges	230
10.3	Other bridges	231
	10.3.1 Transformer bridge	231
	10.3.2 Bridged- and twin-T bridges	233
	10.3.3 Tuttle-type bridges	235
10.4	Practical considerations	240
	10.4.1 Sources and detectors	240
	Additional reading	241
	Examples	241

11	**Circuit theorems**	**243**
11.1	Superposition Theorem	243
11.2	Thévenin's Theorem	245
11.3	Norton's Theorem	248
11.4	Reciprocity Theorem	249
11.5	Compensation Theorem	250
11.6	Star-mesh transformation	251

11.6.1	Star elements in terms of delta elements	252
11.6.2	Delta elements in terms of star elements	253
11.6.3	Realisability of equivalent network	254
11.6.4	Applications	254
	Additional reading	256
	Examples	257

12	**Electrical units**	**258**
12.1	Introduction	258
12.2	Fundamental mechanical units	258
12.3	Systems of electrical units	259
12.3.1	Basic CGS systems	259
12.3.2	Practical and international units	261
12.3.3	Giorgi system	262
12.3.4	Advent of MKSA and SI systems	263
12.3.5	Magnetic constant. Definition of ampere	263
12.4	Establishment of electrical standards	263
12.5	Realisation of ampere	264
12.5.1	Rayleigh current balance	266
12.5.2	Ayrton−Jones current balance	267
12.6	Realisation of ohm	268
12.6.1	Lorenz disc	268
12.6.2	Campbell calculable mutual inductor	270
12.6.3	Calculable capacitor	272
12.7	Material standards	275
12.7.1	Standard resistors	276
12.7.2	Standard cells	276
12.7.3	Zener diode	276
12.7.4	Josephson effect	277
12.7.5	Gyromagnetic ratio of proton	277
12.7.6	Standard capacitors	278
12.7.7	Standard inductors	279
12.7.8	Calibration	279
	Additional reading	280

ANSWERS	281
INDEX	285

Preface

This book has grown out of an annual series of lectures given to under-graduates who include physics as a first year subject at Chelsea College, University of London. The course was conceived as a study of linear electric circuits, to be taught as a self-contained discipline in isolation from field concepts, and in a form acceptable both by intending physicists and by students whose ultimate specialisms lie elsewhere. It was evident at an early stage that no suitable companion textbook would be available, and it is intended that this will remedy the deficiency.

A number of topics have been expanded beyond the customary range of first year studies, and the coverage generally is more extensive than would normally suffice for a one-year course. It is hoped that the book will prove helpful in a variety of courses in physics and electronics leading to first degrees, national certificates and diplomas.

The reader will need to have completed the equivalent of a GCE ordinary level course in physics, and to have followed the study of electric circuits to an appreciably more advanced stage. An elementary knowledge of calculus is required, and familiarity with the complex exponent and with complex algebra generally is essential in the chapters featuring alter-nating current. It has not been considered desirable in an introductory work of this nature to employ the more advanced mathematical tech-niques of the electronicist and electrical engineer. Simplicity in presenta-tion has been a dominant motive, and the quantity of descriptive detail has been limited deliberately. References to general and specialist literature are provided at the ends of those chapters where an interest in further reading is likely to have been stimulated.

Present-day knowledge of the behaviour and potentialities of electric current is a consequence of research and invention spanning a period of about 200 years. The more significant lines of discovery are summarised in a preliminary chapter; in the following chapter it is noted that current and potential difference can be measured with the aid of the various quantitative indicators which form part of the historical legacy, and that scales for these quantities can be established without foreknowledge of the laws of behaviour of the indicators and without the traditional involvement with field concepts. This is the approach formalised by Campbell and Hartshorn, and the writer has found himself quite unable to improve on their exposition. In succeeding chapters the behaviour of representative networks is studied under dc, transient and ac conditions. The reader is encouraged to predict the behaviour of given circuits by recognising characteristic features, rather than by proceeding laboriously from first principles. In the study of ac circuits the complex exponent is introduced at an early stage. The Argand diagram is offered as a useful illustrative aid; phasor diagrams are excluded since these, in the writer's view, are unhelpful and can be conceptually confusing. The ambiguous term *vector* is also omitted. Filters and transmission lines are given more attention than is customary at this level. The treatment of the air transformer is unusually extensive, and includes an investigation of the effects of capacitive tuning of the primary and secondary coils. It is hoped the student will welcome the challenge of a theoretical analysis which is neither unconvincingly approximate nor mathematically too demanding. A detailed account is given of electrical bridges, including transformer and Tuttle types, attention being paid to such practical aspects as screening and optimum component values. In the final chapter, methods of establishing the principal electrical standards are discussed.

Problems are provided at the ends of all chapters where their presence is considered to be an aid to understanding, the numerical answers being collected together at the end of the book. The use of SI is so widespread now as scarcely to constitute a feature of merit; the benefits conferred by any self-consistent system of units are unlikely to be obtrusive in the present restricted context in which, in any case, the ampere, volt and ohm have no serious competitors. The symbols which have been adopted are largely those recommended in 1971 by the Symbols Committee of the Royal Society. The writer has nevertheless favoured the lower case symbols v, i and q for potential difference, current and charge respectively, and has not followed the popular preference for distinctive symbols for complex quantities, on the grounds that these carry the implication of a difference in kind where none exists, and otherwise provide no recognisable advantage.

Acknowledgements

My thanks are due to Kathleen Akerman, who typed the manuscript with her customary care and insight. Dr C. H. Smith provided valuable encouragement during the writing of the book, and corrected errors in the answers provided for the numerical problems. I am indebted to Professor J. E. Houldin for numerous helpful discussions, and for undertaking the task of reading the typescript and identifying many omissions, errors and obscurities.

1

Historical background

1.1 Introduction

Our present-day understanding of the behaviour of electric circuits is based largely on knowledge that has accumulated since the first voltaic cell was assembled at the close of the eighteenth century. The intervening period has been one of continuous research and development, accompanied by the evolution of a vast electrical manufacturing industry. Fortunately, the characteristics of most circuit components are simple, at least to a first approximation, and it is not difficult to design networks capable of a great variety of functions. In this book we shall be concerned with the behaviour of those components which exhibit *linear* electrical characteristics, and as an introduction to this study we shall look first at relevant historical lines of discovery and invention.

1.2 Static electricity

The earliest observed effects of *static electricity* must undoubtedly have been lightning and the aurora borealis. In due course civilised man would have noted the crackling and small sparks produced by combing dry hair, and the attraction of light objects by materials such as amber, following rubbing. A primitive friction machine was devised in 1660. This was the invention of Otto von Guericke, and consisted of a sulphur globe which could be electrified by causing it to rotate while in rubbing contact with a cloth. In succeeding decades increasingly powerful machines were constructed, and a notable sideline to the more serious research investigations was the popular entertainment derived from the displays of static phenomena which they provided.

In 1720 Stephen Gray recognised the distinction between conductors and insulators, and showed that the influences of electric charge could be communicated through distances of several miles along a conducting path of insulated metal wire. Dufay explained repulsion and attraction between accumulations of charge in terms of two kinds of electricity, which he called *vitreous* and *resinous* by association with the substances on which they most commonly appeared. These terms were later changed by Franklin to *positive* and *negative* respectively.

1.3   Magnetism

It is likely that the tendency of fragments of certain rocks to take up a consistent geographical orientation when suitably suspended has been known to man for more than 4000 years, and there is evidence of the occasional early use of this property to assist navigation on land. The magnetisation of iron by rubbing it with fragments of the magnetic ore lodestone probably dates from about AD 1000, and the widespread use of the compass at sea is believed to have developed from about that time.

On rare occasions the properties of a compass needle were found to have been modified by a lightning stroke, and a similar event could result in the acquisition of magnetic properties by iron which was previously unmagnetised. But it was not until the eighteenth century that any systematic relationship between electricity and magnetism was demonstrated, when Franklin magnetised sewing needles by the discharge of Leyden jars through a coil of wire.

1.4   Capacitance and charge

The Leyden jar was discovered in 1745 as a result of efforts to minimise loss of static charge from charged bodies by leakage. In one experiment a nail was driven through the cork of a bottle containing water, and was then connected briefly to the terminal of a friction machine. The intention was that charge would flow from the machine to the water in its insulating container, and to this extent the experiment went according to plan. But when the experimenter accidentally touched the nail with one hand, while still holding the bottle with the other, an electric shock was experienced of far greater severity than any obtainable directly from the machine. We recognise today that this phenomenon results from an electrical property which we call *capacitance*. Although quite small in magnitude for isolated conductors, it is relatively large for the Leyden jar. The essential features of the jar are the insulated conducting coatings on the inner and outer surfaces, which act as electrodes and are able to store large quantities of static electricity. The amount taken up when the device is connected to the electrical machine is proportional to the capacitance, and since this is large a correspondingly severe shock will be caused when the jar is discharged through the human body. The terms *condenser*, and more recently *capacitor* have since come to be used for charge-storing devices. The function of the water in the jar is merely to provide a conducting path from the nail to the inner surface.

By 1749 Franklin had recognised the phenomena of the collection and discharge of electric charge at points, and three years later conclusively demonstrated the electrical character of lightning by flying a kite in a thunderstorm and collecting in a Leyden jar the charge conducted down the wet string. But quantitative measurements of static electricity developed with the utmost slowness. Cavendish judged the electrical condition of a condenser by the severity of the shock which he experienced when he discharged it through his body, and by the width of gap which a spark derived from the condenser could cross. He was admittedly well aware that the two methods provided differing information.

In 1784–5 Coulomb demonstrated the inverse square law of force between electric charges, using a torsion balance to measure the force between two small charged spheres. Knowledge of the law of force between charges enabled a *unit of charge* to be defined, so that a quantitative theory of electrostatics could be established. *Electric potential* could now be defined as the work done in bringing up unit charge to a given location from infinity. Electrical capacitance could in turn be defined for a conductor as the ratio of the total charge residing on it to its potential.

1.5 Electrochemistry

In 1762 Sulzer noticed sensations of taste and itching when two pieces of different metals — silver and zinc — made contact in his mouth. The significance of this discovery was not appreciated until after the observation by Galvani in 1786 of the twitching of frogs' legs which were hung on a copper hook attached to an iron railing. This occurred whenever the free end of a leg was blown by the wind into contact with the railing. Galvani recognised similarities with effects obtained using friction machines, and correctly deduced that the cause was electrical in character. Yet despite mounting evidence to the contrary, he insisted that the source of current resided in the tissues of the frog rather than in the junction of dissimilar metals. This view was at first shared by Volta, who rediscovered Sulzer's phenomenon, and went on to assemble the first battery (1800). This consisted of discs of alternate metals arranged in pairs, each pair being separated by cardboard or leather moistened with saline solution. The voltage generated was sufficiently high to cause electric shocks and produce sparks.

In 1797 Pearson had decomposed water into oxygen and hydrogen by electrolysis, using current drawn from Leyden jars which had been charged from a friction machine. Continuous and much heavier currents now became available using improved forms of the voltaic pile. From 1801 onwards Davy was employing his newly discovered carbon arcs at the Royal Institution as sources of intense light and for the melting of metals. One of his more spectacular successes was the isolation for the first time of the elements sodium and potassium. The hydroxides of these metals do not conduct electricity, and electrolysis of the aqueous solutions results merely in the decomposition of the water content. In 1807 Davy successfully electrolysed the hydroxides in the molten state, producing in each case glowing globules of the metal at the negative electrode. In 1808 he installed an enormous battery consisting of 2000 pairs of zinc and copper plates in an aqueous mixture of nitric and sulphuric acids.

Davy's work at the Royal Institution was continued and extended by Faraday. The latter's laws of electrolysis date from 1832, and with present-day wording state that the mass of a substance liberated in electrolysis is proportional to its chemical equivalent and to the total electric charge which passes.

For some years chemical cells had been the only continuous source of electrical energy. There were problems of local action and polarisation, so that in the interests of economy means had to be provided for lifting the plates out of the electrolyte when the cell was not in use. Local action is caused by local current flow initiated by impurities in the zinc, and the effect was eventually cured in 1828 by amalgamation of the zinc plates with mercury. Polarisation is associated with evolution of bubbles of hydrogen gas at the positive electrode, and was finally eliminated by the use of new designs of cells. In the Daniell cell of 1836 the use of two electrolytes separated by a porous pot gave a continuous ion-exchange process, metallic copper being evolved in place of gaseous hydrogen. The Leclanché cell of 1868 used solid manganese dioxide as a depolariser, and forms the basis of the present-day dry battery, in which each cell generates about 1.5 V.

In 1863 the Daniell cell was adopted as the practical standard of emf, but was displaced from this position ten years later by the Clark cell. This was succeeded in turn in 1892 by the Weston cell, which has retained the status of standard cell to the present day.

1.6   Electromagnetics

There is some evidence that several years passed following the invention of the chemical cell before it was realised that magnetic effects are exhibited by a circuit only when it is closed. It was certainly not until 1819 that

Oersted, working in Copenhagen, noted the response of a compass needle to the closure of a circuit, and subsequently explored the distribution of the field of magnetic force in the vicinity of a current-carrying wire. Meanwhile Arago in Paris observed the magnetisation of a needle situated inside a coil connected to a battery. Within two years of the announcement of Oersted's results, Ampere had completed his brilliant investigation of the magnetic effects of electric circuits, and elucidated the laws of force between current-carrying wires.

1.7   Electrical measurements

In 1820 Schweigger produced his *multiplier*, later known as a *galvanometer*, in which the response of a compass needle to the flow of current in a circuit was greatly enhanced by placing it at the centre of a rectangular coil consisting of 100 turns of wire connected in series with the circuit. This inspired the development of a variety of current-detecting and -measuring devices of low inertia. Just how great was the need for these may be judged by the fact that as late as 1832 the telegraph systems were still using electrolytic effects for the detection of current flow. Precision current measurements became possible in 1837 with the arrival of Pouillet's tangent galvanometer. Weber's electrodynamometer of 1841 utilised the mechanical couple between two current-carrying coils, and later modifications could be used for the measurement of either dc or ac current, and also power.

The first measuring circuit of recognisably modern form was the Poggendorff potentiometer of 1841. This was essentially a metre bridge, which was calibrated by means of a standard cell and could be used to measure potential difference in terms of distance along a uniform wire. The Wheatstone bridge for the comparison of resistances followed in 1843. At this early stage there was still no accepted standard of resistance, and one of Wheatstone's resistance boxes is marked in 'miles'. In fact up to 1850 all units of resistance were based on the size and weight of lengths of conducting wire.

It is perhaps equally surprising that up to 1882 no truly portable precision electrical instrument had been developed. The d'Arsonval galvanometer dates from that year, and was superior in many respects to the earlier moving-iron and hot-wire instruments. It utilised the couple experienced by a current-carrying coil suspended between the cylindrical pole pieces of a permanent magnet. The Kelvin current balance appeared in 1886. This instrument enabled the forces between pairs of current-carrying coils to be compared with the gravitational force acting on a known counterbalancing mass, but it was cumbersome, and rather slow in response. In 1888 Weston produced a pivoted moving-coil instrument which was more robust than the d'Arsonval galvanometer, embodying a design which is still in use at the present day in the majority of portable ammeters and voltmeters.

1.8   Telegraphy

Some mention of the electric telegraph is appropriate at this point because of the impetus which its development gave to the design of detectors and electrical measuring techniques generally, as well as to the production of improved insulators and conductors.

As early as 1747 messages in the form of electric current of static origin had been transmitted across the River Thames through a wire. The newly discovered continuous electric current was used for telegraphy by Sömmering in 1809. His sending and receiving stations were linked by thirty-five conducting wires, which were connected at the receiving end to glass tubes containing water, each representing a different symbol. Transmission of electric current through a particular wire was revealed by the gas bubbles associated with electrolysis of the water in the corresponding tube. In 1832 a magnetic needle with a 'multiplier' was successfully used by Schilling for the detection of coded signals, and subsequent development of the telegraph system was rapid.

The first undersea cable was laid between Dover and Cape Gris Nez in 1850, but communication was brief, as a fisherman raised the cable in his trawl and cut out a considerable length. In the following year a new improved cable was laid, and the success of submarine telegraphy was assured, although cable-laying continued to be a difficult and complex operation.

1.9   Electromagnetic induction

The earliest recorded evidence of the phenomenon of electromagnetic induction was provided by the apparatus known as Arago's disc (1824). Arago had observed that the oscillations of a magnetic needle are damped when a metal plate is brought near. He therefore rotated a copper disc about its axis in a horizontal plane beneath a suspended compass needle, and found that the needle was deflected in the direction of rotation, and that it began to rotate when the rate of rotation of the disc was increased sufficiently. These effects were unexplained at the time, although we would now recognise that the coupling is in each case associated with eddy currents which are induced in the conducting metal.

One manifestation of electromagnetic induction is *transformer effect.* This was discovered by Faraday in 1831 when he switched on a current in one of two coils wound on a soft-iron ring and detected a momentary flow of current in the other. He next demonstrated that a pulse of current flows in a coil connected in a closed circuit when a bar magnet is thrust into it or is suddenly withdrawn. His third major success within the space of two months was the invention which has come to be known as Faraday's disc. A copper disc was rotated with its plane at right angles to the field of several permanent magnets. When Faraday connected a galvanometer between the axle and periphery of the disc the instrument was deflected, showing that a steady current was flowing. This is another manifestation of electromagnetic induction, which we shall call *dynamo effect.*

The Faraday disc provided continuous electric current by the expenditure of human mechanical effort, and for some years many experimental dynamos were similarly hand-operated. Alternative versions were designed to produce alternating current, and in 1836 the first rotating non-mercury commutator was introduced for the conversion of the ac output to dc.

1.10   Electric lighting and mains supplies

An immediate application for power-driven dynamos existed in lighthouses, where arc lights could now be operated using coal-burning steam engines as the source of power. Up to 1880 arc lamps were the only source of electric light anywhere, and at that time docks, theatres, shops and principal streets were being lit in this way. Despite considerable ingenuity in design there remained the problem of the rapid burning away of the carbon electrodes, with a resulting need for frequent servicing. The light output of arc lamps was in any case so intense as to be suitable only for lighting large areas.

Arc lighting became virtually obsolete with the successful demonstration in 1878 of the filament or incandescent lamp. Filament lamps were developed independently by Swan at Newcastle and Edison in the United States, and disputes over patent rights were resolved by the formation of a joint company for large-scale production. Filaments were made of carbon until 1908, when the first tungsten lamps appeared.

Apart from its considerable social impact, electric lighting provided an enormous stimulus to the proliferation of electrical supplies. In 1880 high-power generators were appearing, and by 1885 the largest were capable of delivering 0.1 MW. At this time a preference for dc operation was emerging, as storage batteries could replace the dynamos at night and during breakdowns. Alternating current did not begin to find more general favour for electric power distribution until about 1895.

1.11   Electrical units

As early as 1850 confusion was being caused by uncertainties and inconsistencies in sizes of electrical units. In 1861 a Standards Committee was set up under the chairmanship of W. Thomson (later Lord Kelvin) with a view to remedying this situation, and to making recommendations concerning the closely related area of electrical measurements. In the following year a resolution was adopted that measurements of resistance, current and voltage should be based on a centimetre—gram—second electromagnetic system. The size of the unit of resistance arrived at in this way was inconvenient for the principal applications of the time, and a practical standard was established equal to 10^9 centimetre—gram—second electromagnetic units (CGS EMU). This eventually came to be known as the OHM. The practical units of current and voltage were subsequently chosen to be 10^{-1} and 10^8 respectively times the sizes of the corresponding units in the CGS EMU system.

At the International Electrical Exposition held in Paris in 1881 the names volt, ampere and ohm were formally adopted for the three basic circuit quantities. In 1901 Giorgi formulated a unified metre—kilogram—second (MKS) system of mechanical and electrical units, but nearly half a century was to pass before this was to gain international recognition.

Additional reading

DUNSHEATH, PERCY. *A History of Electrical Engineering*, Faber and Faber, 1969.

TAYLOR, L. W. *Physics. The Pioneer Science.* Vol. II. Light, Electricity, Dover, 1941.

Inventor and Entrepreneur: Recollections of Werner von Siemens, Lund Humphries, 1966.

HOLTON, G. J. and ROLLER, D. H. *Foundations of Modern Physical Science*, Addison-Wesley, 1958.

2

Basic electrical quantities

2.1 Electric current

A continuous electric current can be maintained in a closed conducting circuit if a suitable source such as a battery or dynamo is connected in the circuit. The presence of current can be confirmed by the observation of a variety of phenomena, such as a rise in temperature of some of the components, or the chemical effects of electrolytic action, or the development of electromechanical forces. The heating effect can cause the melting of fuse wire, and may be sufficiently intense to produce incandescence as in an electric filament lamp. Electromechanical forces are used in galvanometers for the sensitive detection of current flow, and where higher powers are available their action is evident in the movement of the cone of a loudspeaker, or the rotation of the armature of an electric motor. As soon as the circuit is broken, as for example when a switch is opened, the various effects cease, indicating that the current is no longer flowing.

Of course, all these observations are merely qualitative in character, and more elaborate experiments are required if one is to establish the precise laws of behaviour of electric current.

2.1.1 Current flow indicators

We shall use the description *indicating instrument* for any device which provides quantitative indications related to the flow of current. There are numerous examples in most research and teaching laboratories. A common form is the moving-coil instrument, and in the present discussion we could include such varied apparatus as chemical voltameters, and moving-iron and hot-wire instruments.

Let several of these instruments be connected in series as in Fig. 2.1, so that a simple closed circuit is formed with the battery and a variable resistor. It would be noticed that no instrument gives any indication of

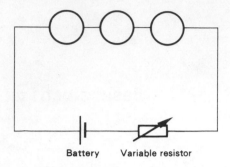

Battery   Variable resistor

Fig. 2.1 Series connection of indicating instruments

flow until the last connection in the circuit is completed, and that in general all the instruments then show some response. If the resistor is varied, the indications of the instruments increase or decrease together.

The response of any instrument is found to be independent of its position in the circuit, provided the sense of connection of its terminals is consistent. Moreover, a given indication by any one instrument is always accompanied by the same set of indications by the others. The reversal of some instruments gives rise to a change in their indications, while others are unaffected by reversal. Instruments of the first type will be called *direction-sensitive*, and those of the second will be called *reversible*.

The above observations suggest a similarity to the flow of water in a pipe. The battery is analogous to the pump, the resistor to a valve, and the closed conducting path to the pipe. The indicating instruments are rate-of-flow meters of various kinds. The analogy suggests very strongly that if a more elaborate circuit is constructed, in which several branches lead into a main conductor, the characteristic feature 'current' of the main circuit may be the sum of the corresponding features of the branches. It suggests in other words that there is a property 'current', which is characteristic of electric circuits, and which obeys a law of addition in branched circuits.

2.1.2   Establishment of scale of current

Let circuits X and Y be arranged as shown in Fig. 2.2, with a common limb in which are connected several indicating instruments such as A_1 and A_2. It is convenient to refer to the instruments as if each possesses a pointer moving over an initially blank scale. The method of indication would of course be much less direct for a device such as a chemical voltameter.

Notice also the instruments U and B. The former is arranged to

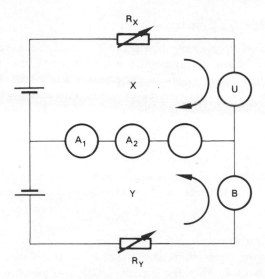

Fig. 2.2 Circuit for establishment of scale of current

define a 'unit' of current, by marking a figure '1' at some arbitrarily chosen point on its scale. The scale of instrument B is left blank for the moment. All the instruments are set to indicate zero readings in the absence of current flow. When circuits X and Y are completed it is neces- sary to reverse the connections of any instrument which is giving negative indications.

Circuit Y is broken, and the resistor R_X is adjusted until the pointer of U is at the position '1'. The scales of A_1 and A_2 are marked with the figure '1' at the points corresponding to their indications. Circuit X is now broken and Y is closed. The resistor R_Y is adjusted until the pointers of A_1 and A_2 are again at the position '1'. The scale of B is marked with a '1'. Circuit X is closed, and the resistors are adjusted until both U and B indi- cate '1'. The indications of A_1 and A_2 are now marked with a '2'. Proceed- ing in the same way, circuit Y is used to enable the corresponding indication of B to be marked with the numeral '2'. Then with B at '2', and U in the condition '1', A_1 and A_2 are marked '3', and so on.

The process is continued to progressively higher readings, until it is considered that the scales have been calibrated to a sufficient extent. Each of the instruments can now be regarded as providing a quantitative measure of current. It would be noticeable at this stage that the scales of many direction-sensitive instruments would have been subdivided accord- ing to an approximately linear law; the scales of many reversible instru- ments would be found to have been subdivided according to an approximately square law.

2.1.3   Kirchhoff's first law

The above procedure gives rise to a scale of current which is based on the arbitrarily selected unit '1', and which is quite independent of the laws of operation of the indicating devices. The establishment of the scale depends on the assumption of a restricted law of addition for currents flowing from a junction into two branches of a circuit, of the form

$$m'1' + '1' = (m + 1)'1',$$

where m is a positive integer.

By substituting a calibrated instrument such as A_1 in place of U, it is easy to demonstrate the validity of the more general law of addition

$$m'1' + n'1' = (m + n)'1',$$

where n is a positive integer.

Now let the connections of the instrument B and of the battery in circuit Y be reversed. When the currents in the two branches are of equal magnitude the current in the centre limb is observed to be zero. The conclusion is drawn that electric current is a directed quantity, and that the sign associated with its magnitude must be reversed when its direction is reversed.

We therefore have for a junction in a circuit

$$\Sigma \text{ currents arriving} = \Sigma \text{ currents leaving.}$$

It is convenient to attach a positive sign to currents arriving at a junction, and a negative sign to currents leaving. The above relation then takes the form

$$\Sigma \text{ currents arriving at a junction} = 0.$$

This is Kirchhoff's first law for electric circuits. In the steady state it is equally valid when the concept of a junction is extended to include an entire electrical system, however complex this might be.

2.1.4   Electric charge

Faraday's experiments showed that the mass m of an element which is transported or liberated in an electrolytic process is proportional to the steady electric current i, and to the time t of flow. This result is expressed by the relation

$$m = zit,$$

where z is a constant for the element, and is known as the electrochemical equivalent.

We have seen that electric current has the character of a rate of flow. The quantity '*i* times *t*' can be interpreted as a *total flow* occurring in the time interval *t*. In electrolysis it is identified with the amount of charge conducted between the electrodes of an electrolytic cell, and the above relation indicates that it is proportional to the mass of an element which is simultaneously transported through the electrolyte. We shall therefore recognise '*i* times *t*' as possessing the nature of charge, and will represent it by the symbol *q*, so that

$$q = it.$$

2.1.5   Conservation of charge

Where the current is time-dependent, the more general relation

$$q = \int_{t_1}^{t_2} i \, \mathrm{d}t$$

is employed, so that

$$i = \frac{\mathrm{d}q}{\mathrm{d}t}.$$

Electric current may in consequence be represented as a rate of passage of charge past a point in a circuit. Kirchhoff's first law is seen to be equivalent to the statement that the net rate of arrival of charge at a junction in a circuit is zero, and it therefore expresses a Principle of Conservation of Charge.

2.2   Establishment of scale of voltage

In addition to those instruments finding application as current-measuring devices, a second group may be identified whose members greatly impede the flow of current when inserted *in series* with a circuit, but give nevertheless definite indications when connected *across* any component of a circuit. This second group includes electrostatic instruments, potentiometers and high resistance galvanometers. Potentiometers contain batteries, and are in consequence direction-sensitive. Some members of the group are reversible. All can be used to establish the same magnitude, which we shall call voltage. Instruments of this type are called voltmeters.

A law of addition can be established for voltages, similar to that for currents. In Fig. 2.3 current supplied by the battery flows through a resistor. The voltmeter U is marked with the reading '1' when connected to the resistor at points X and Y. Then a second meter A, which is

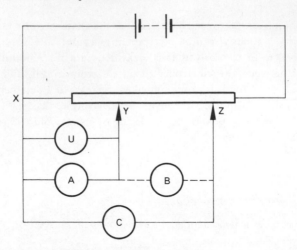

Fig. 2.3 Circuit for establishment of scale of voltage

connected at the same points, is also marked '1'. It is now transferred to position B, and the positions of sliders Y and Z are adjusted until both meters again read unity. A third meter C, which is connected as shown, is now marked '2'.

Meter A is now connected in parallel with meter C, and any necessary adjustment of slider Z carried out so that the latter instrument again reads '2'. Meter A is now marked '2'.

Continuing the process, a voltage scale can be marked on each of the three instruments. They can then be employed to verify the law of addition by using various combinations of readings.

It will be apparent on reflection that in the methods of calibration which have been described here a linear addition law has been presupposed both for current and voltage. For each the reading '2' was selected as a sum of readings '1' plus '1'. It could have been called '1½' instead of '2', so deliberately excluding the possibility of a simple law of addition. However, having marked the instruments in the chosen way, the validity of the calibration could be confirmed by showing that consistent results are obtained for all addition processes, e.g. '4' + '3' = '7'.

In experiments such as the above, voltages of both signs would be encountered. In circuit analysis one must often be very careful in deciding what sign should be attached to a voltmeter reading. Suppose that a resistor is connected between two points A and B in a network. If current flows through the resistor from A to B, then A is said to be positive with respect to B, and voltmeter readings v_{AB} and v_{BA}, although equal in magnitude, are taken as positive and negative respectively.

It may be wondered how it has been possible to identify the direc-

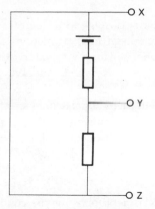

Fig. 2.4 Demonstration of summation law for voltages

tion of flow of current. The answer is simply that in the earliest experiments involving electrolysis and motor effects, one sense of flow was arbitrarily defined as positive, and the other as negative. This convention has been retained to the present day, and is perpetuated in the labelling of the terminals of all moving-coil instruments.

In Fig. 2.4, a battery is connected to form a closed circuit with two resistors. No reading would be observed on a voltmeter connected between terminals X and Z, while equal finite readings would be produced on connecting successively at XY and YZ, provided the voltmeter terminals are interchanged in these last two situations. We have therefore for the sum of these voltmeter readings the relation:

$$v_{XY} + v_{YZ} = 0.$$

The circuit could be subdivided further, and the sum of the observed voltages would still be found to be zero, provided due regard is paid to signs. This result can be expressed by the simple equation

$$\Sigma v = 0.$$

This important conclusion will be given further attention in section 2.5.

2.3 Resistance

2.3.1 *Ohm's law*

The procedures outlined above provide instruments calibrated for the measurement of current and voltage on arbitrary scales. These could be used to investigate the relation between current and voltage for a conductor. We shall use the term *potential difference* for the voltage which develops across a conductor through which current is flowing, although its use is wider than this statement implies (section 2.4). For a large number of conductors strict proportionality is found to exist between current and voltage, provided these are steady and not excessive in magnitude, and provided the temperature of the conductor remains constant. This is Ohm's law, which is obeyed by a great number of substances but which fails in particular for gases and semiconductors.

Thus for a given conductor obeying Ohm's law, the relation

$$\frac{v}{i} = R$$

applies. The constant R is a property of the conductor called the *resistance*. The circuit element itself is called a *resistor*.

2.3.2   *Establishment of scale of resistance*

A scale of resistance can be established as follows. Equality of two resistances is demonstrated by showing that the current in a circuit containing one is unchanged by substitution of the other. The two are now connected in series, and a new resistor is selected which can be substituted for the two, again without change of current. Continuing in this way, a set of resistors can be assembled with magnitudes differing by unity on an arbitrary scale.

Notice that the resulting resistance scale is produced independently of other electrical properties. Any bridge or balancing circuit which is employed for the comparison is acting as a detecting device, rather than as a measuring instrument.

The unit of resistance on this scale depends on the values of the first two resistances selected. Strictly, the relation

$$v = iR$$

should read

$$v = KiR,$$

where K is a constant and is not necessarily dimensionless. But at this point we shall anticipate some later discussion (section 2.7.3) and assume the unit of resistance has been selected to make K equal simply to unity.

2.4   Electromotive force

There is an additional class of circuit elements which do not obey Ohm's law, and which are found to display a finite terminal voltage even when

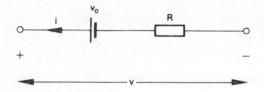

Fig. 2.5 Equivalent circuit of linear source of emf

passing zero current. This group includes batteries, dynamos, thermo-
couples and potentiometers. For such devices the relation between current
and voltage takes the approximate form

$$v = v_0 - iR,$$

where v_0 and R are constants. Examination of this relation shows that v_0
can be identified as the open-circuit voltage. R is called the *internal resistance*. The element can be represented by the simple circuit of Fig. 2.5. v_0
is called the *internal electromotive force*. In general *electromotive force*, or
emf, is the open-circuit voltage developed by an active device such as a
battery or dynamo, and is capable of initiating continuous current flow if
a suitable closed circuit is provided. *Potential difference* is a more general
term. It is used for the voltage developed across an active or passive device.
For a resistor its magnitude falls immediately to zero when for any reason
the flow of current ceases.

2.5   Kirchhoff's second law

The relation

$$\Sigma v = 0$$

which was established in section 2.2 is valid for any closed path traced out
in a circuit, and will be taken as a general statement of Kirchhoff's second
law for electric circuits. Suppose now that a number of pure resistors and
sources of emf are connected in a closed series circuit. The voltages
developed across the various components must total zero. It follows that

$$\Sigma (v_0 - iR_s) - \Sigma iR_j = 0,$$

where i is the current flowing, the iR_j terms are the potential differences
developed across the pure resistors, and the $v_0 - iR_s$ terms are the voltages
developed across the sources of emf. Therefore

$$\Sigma iR_s + \Sigma iR_j = \Sigma v_0,$$

which can be written more simply as

$$\Sigma iR = \Sigma v_0,$$

where the summation on the left-hand side extends over resistances of all kinds in the circuit.

2.6   Energy and power

Any source of emf requires a supply of energy to enable it to maintain a flow of current. This energy may for example be thermal as in a thermocouple, or chemical as in a battery, and it can subsequently reappear in the same or alternative forms during the period of current flow.

If the energy is produced in thermal form, it is relatively easy to measure the amount involved using calorimetric techniques, although the accuracy involved is not very high. In principle it is only necessary to connect in series with the circuit a resistance in the form of a heating coil, and to immerse this in a known mass of water. The temperature rise resulting from a flow of steady current over a noted time interval is then measured.

Experiments like this show that the energy W produced in a resistor is proportional to the product of the three quantities voltage, current and time, or in terms of symbols:

$$W \propto vit.$$

If the units of these quantities are chosen to make the constant of proportionality unity, then the relation becomes

$$W = vit.$$

This may be written alternatively

$$W = vq.$$

The rate of supply of energy is called the *electric power P*. This is the time-derivative of W, so that

$$P = vi.$$

For a conductor obeying Ohm's law, the relation

$$v = iR$$

holds. We have therefore for the electric power in a resistance R the equivalent relations

$$P = vi = i^2 R = v^2/R.$$

An expression for the energy dissipated in time t by a steady flow of current can be obtained by multiplying any of these quantities by the factor t.

2.6.1   *Conservation of energy*

We have seen that for a resistor the quantity vi is the rate of *conversion* of electrical energy to alternative forms. It is reasonable to infer that for a source of emf v_0 carrying current i, the quantity $v_0 i$ is the rate of *delivery* of electrical energy into the circuit by the source.

Let Kirchhoff's second law be applied to a closed series circuit consisting of resistances and sources of emf. Then

$$\Sigma \, iR = \Sigma \, v_0.$$

The current i is common to all components in the circuit. Let us multiply each term in the equation by the factor it, where t is an interval of time. Thus

$$\Sigma \, i^2Rt = \Sigma \, v_0 it.$$

The terms on the left-hand side are the electrical energies dissipated in the various resistors. The terms on the right-hand side are the electrical energies delivered into the circuit by each source of emf in the same time interval. The Law of Conservation of Energy requires that the two sides of the equation should be equal, and Kirchhoff's second law is therefore consistent with this law.

2.7   Electrical units

The laws of behaviour of electric circuits do not depend on the sizes of the units chosen for current and voltage. International agreement on these sizes is nevertheless essential, so that measurements in one location with a particular collection of apparatus can be related to measurements made with a different collection at any other location.

The subject of units is dealt with at some length in Chapter 12. Quite a lot of the topics discussed there can be better understood after careful reading of some of the earlier chapters in this book. It will however be necessary to define some of the electrical units now. An elementary introduction is therefore given below, with principal definitions.

The acceleration $\ddot{x}$ produced in a mass m is related to the applied constant force F by the equation

$$F = m\ddot{x}.$$

The dimensions of force are therefore MLT^{-2}. The units of mass and length will be taken as the kilogram and metre respectively, standards of which are kept at the International Bureau of Weights and Measures at Sèvres, in France. The unit of time is the second, which approximates closely to 1/86 400 of the mean solar day. The unit of force in this system is called the newton.

The work done by a constant force F in moving its point of application through a distance x is given by the relation

$$W = Fx.$$

The dimensions of work are therefore ML^2T^{-2}. The unit of work, or energy, is called the joule, and its magnitude is defined by this equation as part of the universally accepted mechanical system of units.

The equation

$$W = vit$$

relates the steady current i in a circuit component and the voltage v developed across it to the energy W delivered into the component in time t. We have already seen in section 2.6 that the equality of the two sides of this equation is dependent on the use of a suitable system of units for the various quantities involved. The units selected for energy and time are the joule and second respectively, so that a freedom of choice exists for the size of the unit of current or of voltage, but not for both. The establishment of a fundamental and accurately reproducible standard has been very much more difficult for voltage than for current. The accepted procedure at the present time is therefore to define the unit of current first, and then to relate to it the fundamental unit of voltage.

2.7.1 Unit of current

The definition of the unit of current could in principle involve any of the measurable effects of current flow. It is clearly desirable that the magnitude of the unit so chosen should not depend on the properties of any particular substance, and this consideration excludes definitions based on such effects as electrolysis and heating.

The internationally accepted unit of current is the *ampere*, which is defined as that unvarying current which, when maintained in two parallel infinitely long rectilinear conductors of negligible circular section placed at a distance of one metre apart *in vacuo*, produces between these conductors a force of 2×10^{-7} newtons per metre length.

This definition is seen not to involve the physical properties of any substance. But it would not be possible to use the arrangement as a direct means of establishing the unit of current. On the one hand the force of attraction per metre length is extremely small, and on the other the total force between the infinitely long parallel conductors would be infinitely large!

We shall see in Chapter 12 that the system of mutual forces between current-carrying coils of wire provides a much more realistic means of establishing a scale of current. The forces produced in this way are relatively large, conveniently localised and more easily measured. In order to calibrate a current-indicating instrument, in principle one would connect it in series with the coils, and mark the initially blank scale with the current values deduced from the measured forces. The resulting calibrated instrument could then justifiably be called an ammeter.

2.7.2   Unit of potential difference

The practical unit of potential difference is called the *volt*. It is defined as that potential difference which exists between two points in a circuit when, with a steady current of one ampere flowing, one joule of work is done per second. This definition is consistent with the equation

$$W = vit.$$

Unfortunately, any attempt to use the definition in a direct way to set up a scale of voltage involves the measurement of a quantity of energy. It is difficult to measure energy accurately, and consequently an unacceptable limit is set on the accuracy with which the scale can be established. It is worth noticing at this point that there exist on the one hand *theoretical standards* which define the units of electrical quantities in absolute terms, and on the other *practical standards* which represent their closest realisation. We shall find in Chapter 12 that a practical scale of potential difference which is consistent with the above definition is best arrived at in a rather indirect way.

2.7.3   Unit of resistance

We first encountered electrical resistance (section 2.3.1) via the relation

$$v = iR,$$

and the unit of resistance is most suitably defined by means of this. The practical unit of resistance is called the *ohm*, and is defined as the resistance of a conductor across which a potential difference of one volt exists when a steady current of one ampere is flowing in it. Notice that the form of the definition is such that no dimensionless factor is required in the above equation.

Additional reading

CAMPBELL, N. R. and HARTSHORN, L., *The Experimental Basis of Electromagnetism*. Proc. Phys. Soc. 58, 634 (1946).

Examples

1. What steady current flowing in a silver voltameter causes deposition of 55.9 mg of silver on the cathode in 16 min 40 sec?
    (The electrochemical equivalent of silver is 0.001118 gC^{-1}.)

2. What total charge circulates while an initial current of 3A is reduced linearly to zero in a period of 4 minutes?

3. If the specific heat of water is 4200 J kg^{-1} $(°C)^{-1}$, how long does it take a 2 kW electric heater to boil 10 litres of water, the initial temperature being 40 °C?

4. If electricity costs 1p per kW hr, what is the cost of running a 3 kW electric fire for three hours every day for a week?

5. What current flows in a 1 kΩ resistor in which the power consumption is 40 W? What is the potential difference across the resistor?

6. What are
    (a) the operating current, and
    (b) the operating resistance of a 200 V, 50 W lamp?
    What is the amount of energy consumed by the lamp in one hour, and what quantity of charge passes in the same period?

7. What current is drawn from a dc generator supplying 0.25 MW at a pressure of 1 kV? If the output is converted with 90 per cent efficiency to 450 V, what current is supplied at this lower voltage if the generator continues to deliver 0.25 MW?

3

Steady current circuit analysis

3.1 Resistance

3.1.1 Resistances in parallel

In the arrangement of Fig. 3.1, several resistors are shown connected in parallel, providing alternative paths for the current i. The plus and minus signs indicate the polarity of the potential difference v, and are seen to be consistent with the direction of flow of the current.

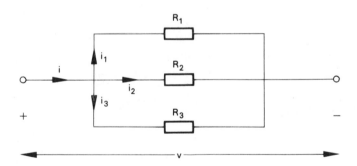

Fig. 3.1 Parallel connection of resistors

Applying Kirchhoff's first law to the left-hand common junction of the resistors

$$i = i_1 + i_2 + i_3.$$

Dividing through by v,

$$\frac{i}{v} = \frac{i_1}{v} + \frac{i_2}{v} + \frac{i_3}{v}.$$

If the effective resistance of the arrangement is represented by the symbol R, then

$$\frac{i}{v} = \frac{1}{R},$$

and

$$\frac{1}{R} = \frac{1}{R_1} + \frac{1}{R_2} + \frac{1}{R_3}.$$

Thus for resistances R_1, R_2, ... R_j, ..., R_n connected in parallel, the effective resistance R is given by

$$\frac{1}{R} = \sum_1^n \frac{1}{R_j}.$$

For parallel arrangements, it is often more convenient to work in terms of a related quantity called *conductance*. This is defined as the current flowing in a conductor divided by the potential difference developed across it. It is represented by the symbol G, and for a given conductor is evidently equal to the reciprocal of the resistance. For any number n of conductances connected in parallel, the effective conductance is given by

$$G = \sum_1^n G_j,$$

where G_j is the conductance of the jth branch.

The unit of electrical conductance is the *siemens*.

3.1.2   Resistances in series

Series connection of resistors is illustrated in Fig. 3.2. It is now the current which is common to each resistor, and the applied potential difference which is distributed between them. This is the reverse of the situation in the previous section for resistors connected in parallel.

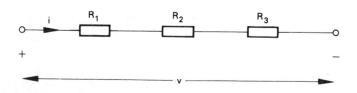

Fig. 3.2 Series connection of resistors

By Kirchhoff's second law, the total potential fall is equal to the sum of the potential differences across the individual components, so that

$$v = iR_1 + iR_2 + iR_3 = i \sum_1^n R_j,$$

where the summation extends over all the resistances. Defining

$$R = \frac{v}{i}$$

as the effective resistance, it follows generally that

$$R = \sum_1^n R_j,$$

this result being valid for any number n of resistances connected in series.

3.1.3   Resistivity and conductivity

Imagine a conductor in the shape of a rod of uniform cross-section, carrying current which within the material is everywhere flowing parallel to the length of the rod, and is uniformly distributed across the cross-section.

If two identical such conductors are connected end-to-end in series, the summation law for series resistances indicates that the total resistance is twice the resistance of either. The length of the composite arrangement is double the length of each conductor, so the resistance of a uniform conductor evidently varies in proportion to the length L of the path of the current.

If alternatively the two identical conductors are connected side-by-side as a parallel arrangement, the relation

$$\frac{1}{R} = \sum_1^n \frac{1}{R_j}$$

indicates that the resistance is precisely halved. The effective cross-section has been doubled by the addition of the second conductor, which indicates that the resistance of a conductor varies inversely as the cross-sectional area A.

These conclusions can be combined to form the single relation

$$R = \rho L/A,$$

where ρ is independent of the length and cross-sectional area, and is in fact a property of the material of which the conductor is composed. It is called the *resistivity*, or *specific resistance*. The units are ohm metre.

By inverting both sides of the equation, we obtain an expression for the conductance G in the form

$$G = \sigma A / L,$$

where the quantity σ replaces $1/\rho$. σ is called the *conductivity* or *specific conductance* of the material. The units are ohm^{-1} metre^{-1} or siemens per metre.

For a cube of side one metre, in which the direction of current flow is perpendicular to a pair of opposite faces, it can be seen that R and ρ are numerically equal, as also are G and σ. The units of ρ and σ are in consequence sometimes given rather confusingly as ohm per metre cube and siemens per metre cube respectively.

The distribution of the current in a conductor can depend markedly on the geometry and positioning of the electrodes, so that these factors influence the resistance presented by the sample to the source. If a resistor obeys Ohm's law, its material is said to be *ohmic* in behaviour. For non-ohmic materials the resistivity and conductivity depend on the magnitude of the current. Even for ohmic materials these quantities are likely to be affected by the temperature change caused by the joule-heating due to current flow. For electrically anisotropic materials, the resistivity and conductivity vary with direction, so that the electrical behaviour cannot be described in terms of simple scalars like ρ and σ. These must then be replaced by quantities called tensors, and the resistance depends in a correspondingly more complicated way on the geometries of the sample and electrodes, and on the positioning of the latter.

3.2   Symbols and sign conventions

For the diagram of Fig. 3.3, the current i will be taken to be positive if it is flowing in the direction indicated by the arrow, and negative if the flow is

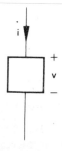

Fig. 3.3 Current flow in a circuit component

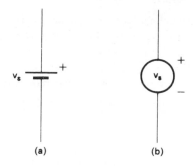

(a)   (b)

Fig. 3.4 Alternative representations of source of constant voltage

in the reverse direction. With this representation, i and v have the same sign for a resistor. If in the course of analysis a current is found to be negative, it is best to let the algebra take care of the sign and to leave the diagram unchanged. Similarly, in the same diagram the potential difference v is numerically positive if the polarity is as shown, and is otherwise negative.

A source of constant voltage will be represented in either of the alternative ways shown in Fig. 3.4. The left-hand diagram is the more apt for dc circuits. The potential difference between the terminals is taken to be independent of the magnitude and direction of the current. For the purposes of analysis it is often convenient to suppose that v_s is continuously variable. If it is reduced to zero the device then has the character of a short-circuit, since it may still carry current. To a good approximation well-maintained lead accumulators, many dynamos and any active device provided with automatic stabilisation of the output voltage can be regarded as constant-voltage sources. They all possess very low internal resistance.

A source of constant current (Fig. 3.5) has a prescribed current flowing through it which is unaffected by the magnitude of any voltage that develops across it. If it is supposed that the source current is variable, and that it can be reduced to zero, the device will then behave electrically as

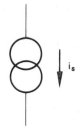

Fig. 3.5 Constant current source

an open circuit. Any voltage source can be made to approximate to constant-current operation by connecting a high resistance in series with it. The current drawn from such an arrangement tends to be independent of the resistance of the load, provided this is small compared with the source resistance. A constant-current source constructed in this way is unfortunately likely to be able to provide only a small current.

3.3   Mesh currents

Mesh currents are shown for the simple network of Fig. 3.6. Mesh currents are sometimes called loop or circuital currents and were introduced by Maxwell. Their use ensures automatic satisfaction of Kirchhoff's first law, since in the vicinity of a junction each mesh current makes identical contributions to the sum totals of currents arriving and leaving. In any circuit element flanked by two loops, the *net* or *linear* or *branch current*, as it may be called, is the algebraic sum of the superimposed currents. Thus the branch current in BC is simply i_1, and the currents flowing towards D in the branches AD and BD are $i_2 - i_3$ and $i_3 - i_1$ respectively.

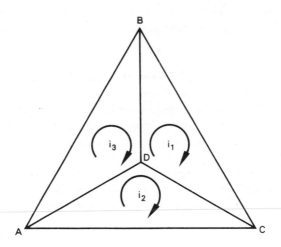

Fig. 3.6 **Example of mesh current representation**

3.4   Sign conventions and Kirchhoff's second law

The relation

$$\Sigma \, iR = \Sigma \, v_0$$

was derived in section 2.5 as an expression of Kirchhoff's second law, and can be used to determine the current distribution in network problems.

Great care is essential in identifying the signs of the various quantities involved.

For the simple circuit of Fig. 3.7, an expression for the current flowing can be obtained without formal recourse to Kirchhoff's second law. It is obvious on inspection of the diagram that the sense of flow must be clockwise, and that the magnitude of the current is given by

$$iR = v_0.$$

Comparison of this result with the more general formula above indicates that where the polarity of a battery of emf v_0 is such as to encourage the flow of a current i, then the emf and current should be provided with like signs.

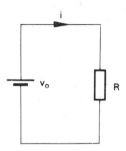

Fig. 3.7 Simple circuit for illustration of Kirchhoff's second law

In more complicated networks, the general formula has to be applied separately to each closed loop, so that each gives rise to a different equation. The following procedure takes account of the rule which we have just established for the signs of the various terms. Select as starting point the positive terminal of any voltage source in the loop under consideration. The emf of this source is to be reckoned positive, as is the emf of any other source in the loop which tends to drive current in the same direction. The emf of any source connected in opposition is then to be reckoned negative. Now trace a path leading away into the circuit, setting off in a direction which does not pass first through the source chosen as starting point. Take as positive those iR terms where the path is followed in the direction of current flow, and negative where the path runs contrary to the direction of current flow.

It may be found helpful if this recommended procedure is first tried out with the simple circuit of Fig. 3.7.

This method of analysis gives rise in general to a set of simultaneous equations which are equal in number to the separately identifiable loops in the network. In many networks simplifying features and symmetries can be identified which reduce the number of equations and the amount of labour subsequently involved.

In order to illustrate the principles laid down here, in the next section we shall solve for the current distribution in several simple circuit arrangements.

3.5   Worked examples

(*a*) Consider the network of Fig. 3.8, in which mesh currents i_1 and i_2 are indicated for the two closed loops. A further loop may be recognised as the outer boundary of the circuit, but no additional information would be contributed by taking account of this, since two simultaneous equations suffice for the determination of the two mesh currents.

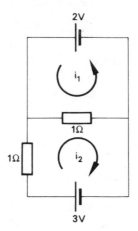

Fig. 3.8

For the upper loop, select as starting point the positive terminal of the 2 V battery, and proceed anticlockwise. The branch current in the upper 1 Ω resistor is $i_1 + i_2$. The battery emf and the iR term are both positive, so that Kirchhoff's second law gives simply

$$(i_1 + i_2) \cdot 1 = 2.$$

For the lower loop, take the positive terminal of the 3 V battery as starting point, and proceed in a clockwise direction. The general formula gives

$$i_2 \cdot 1 + (i_1 + i_2) \cdot 1 = 3.$$

Comparing the two simultaneous equations, it is apparent that

$$i_2 \cdot 1 = 1,$$

and if this is substituted in either equation one obtains

$$i_1 . 1 = 1.$$

The solutions for the mesh currents are therefore

$$i_1 = i_2 = 1\text{A},$$

from which one can solve for all branch currents.

The accuracy of the calculations can be checked by inspecting the circuit to judge whether the numerical results are consistent with the given circuit parameters. The upper 1 Ω resistor carries a total current of 2 A, so that a potential difference of 2 V develops between its ends, the left-hand end being the more positive. This is due to the influence of the 2 V battery. The emf of the 3 V battery is divided as 1 V across the lower 1 Ω resistor and 2 V across the upper.

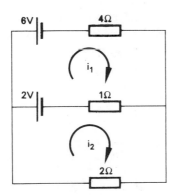

Fig. 3.9

(*b*) The current distribution of the network of Fig. 3.9 will be solved in terms of the two mesh currents i_1 and i_2. Take the positive terminal of the 2 V battery as the starting point, and proceed clockwise for the upper loop, and anticlockwise for the lower loop. The following simultaneous equations are obtained:

$$4i_1 + (i_1 - i_2) . 1 = 2 - 6$$
$$-2i_2 + (i_1 - i_2) . 1 = 2.$$

These are easily solved by standard methods to give

$$i_1 = i_2 = -1 \text{ A}.$$

The equality of the two mesh currents indicates that no current flows in the central branch. The potential difference developed between the extremities of this branch is therefore equal to the emf of the 2 V battery which it contains. This same potential difference appears across the lower and upper branches, and a little reflection will indicate that equal currents of 1 A must flow in the 2 Ω and 4 Ω resistors, as we have already found.

(*c*) In the network of Fig. 3.10, two mesh currents again suffice to represent the current distribution. Select the positive terminal of the 4 V battery as the starting point, and proceed clockwise to set up an equation for the upper loop, and anticlockwise for the lower loop. The two simultaneous equations obtained are:

$$2i_1 + (i_1 - i_2) \cdot 7 = 4 - 6$$

$$-2i_2 + (i_1 - i_2) \cdot 7 = 4 - 2.$$

The solutions are easily shown to be

$$i_1 = i_2 = -1 \text{ A.}$$

As in the previous problem no current flows in the central limb, so that the potential difference between its extremities is 4 V. It is left to the reader to confirm that this result is in accord with the current distribution found for the other branches.

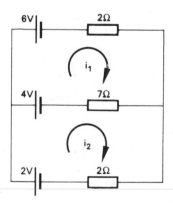

Fig. 3.10

An additional consequence of the absence of current in the central limb is that the current and potential distributions should be unaffected by its resistance, or even by its complete removal. The network is then reduced to a single loop, in which a net emf of 4 V drives a current of 1 A in an anticlockwise direction through a total circuit resistance of 4 Ω.

3.6   Maximum power theorem

The ideals of constant-current and constant-voltage operation can only be

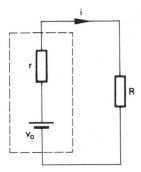

Fig. 3.11 Resistive source with load R connected

approximated in practice, and all real sources behave as if possessing finite resistance. In Fig. 3.11, a dc supply is represented as a source of emf v_0 in series with an ohmic resistor r. Suppose now that a variable resistor R is connected as a load across the terminals of the device. The current flowing is

$$i = v_0/(R + r),$$

and the power consumed by R is

$$P = i^2 R = v_0^2 R/(R + r)^2.$$

The condition for maximum power can be obtained by equating to zero the differential coefficient of P with respect to R. Then

$$\frac{dP}{dR} = v_0^2 \frac{(R + r)^2 \cdot 1 - R \cdot 2(R + r)}{(R + r)^4} = 0,$$

and it is easy to extract from this the condition

$$R = r.$$

This result is expressed in the *maximum power theorem*, which states that maximum power is delivered into a load when the load resistance is equal to the internal resistance of the source.

The maximum power which can be extracted from the source is evidently

$$P_{max} = v_0^2/4R.$$

In this condition half of the electrical power developed by the source is dissipated in the internal resistance of the source. For sources such as car batteries the condition of maximum power is therefore without practical significance, but in many situations the theorem is of great importance, and in section 6.12 we shall see that with suitable modification it can be extended to ac sources.

3.7   Ladder networks

The network of Fig. 3.12 is connected at one end to a source of emf v_0, and will be assumed to extend uniformly away from the source to infinity. The arrangement constitutes a semi-infinite resistive ladder constructed as an endless sequence of series resistors r and shunt resistors r_0.

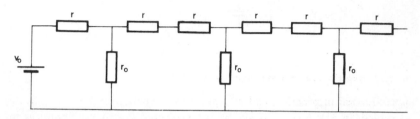

Fig. 3.12 **Resistive ladder network**

If the circuit is imagined to be repeatedly subdivided, a recurring constructional element can be identified. This could take the form of the T-section of Fig. 3.13. Other representations are possible, but this one is particularly convenient for our immediate purpose.

In some ladder networks, resistors are located in both of the series paths of Fig. 3.12, but for purposes of analysis it is more convenient to group all the series resistors in the upper path.

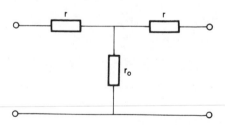

Fig. 3.13 **Representative element of resistive ladder network**

3.7.1   *Matched ladder*

The electrical behaviour of a semi-infinite ladder is unaffected by the removal of any arbitrary number of representative elements from the source end, since the structure that remains is still semi-infinite in length. Let the resistance presented to a source by such a ladder be R. If a resistance of this magnitude is connected as a termination for the severed group of elements, then the composite arrangement must behave electrically in the same way as the original semi-infinite ladder, at least as viewed from the input end, so that the input resistance will also be equal to R.

Let the circuit of Fig. 3.14 represent the T-section in this group which is immediately adjacent to the sever. The appropriate terminating load R is attached, so that the input resistance of the arrangement is equal to R. If additional T-sections are now inserted at the input end, the input resistance will remain unchanged.

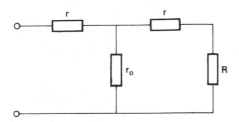

Fig. 3.14   T-section with load R attached

A formula for R will now be obtained. The input resistance of the arrangement of Fig. 3.14 is

$$R = r + r_0 (r + R)/(r_0 + r + R).$$

This is easily reduced to the form

$$R = \sqrt{(r^2 + 2rr_0)}.$$

R is called the *correct* or *matching load*, and it is also equal to the input resistance of the semi-infinite ladder.

The *input resistance* is of course the resistance presented to the source by the ladder with its load connected. The *output* or *internal resistance* of the ladder is the resistance viewed looking back into the ladder from the position of the load, the ladder and its electrical supply being regarded jointly as a voltage source which possesses series resistance. If the internal resistance of the source happens to be equal to the matching resistance R, then because of the symmetry of the ladder its internal resistance will also equal R. This is not the unlikely or special event which it might at first seem. It would be quite common to select approximate equality for the source resistance and ladder matching load, in order to ensure maximum power delivery by the source into the ladder. The same principle reinforces the desirability of matching the load to the internal resistance of the ladder. These considerations tend to limit the choice of source and load for a given ladder network.

3.7.2   *Ladder attenuator*

The discussion which follows refers to the matched ladder. In Fig. 3.15
each of the voltages $v_0, v_1, v_2, \ldots v_n$ is developed across an effective resistance equal in magnitude to the matching load. In Fig. 3.16 a representative

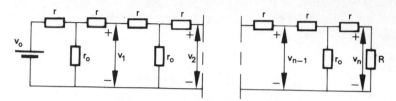

Fig. 3.15 Matched ladder attenuator

T-section is shown with its equivalent load R attached. Remembering that
the effective resistance presented to the voltage v_p is also equal to R, it can
be seen that the input resistance at the terminals of the shunt resistor r_0 is
$R - r$. Let the voltage appearing across this resistor be v' as indicated.

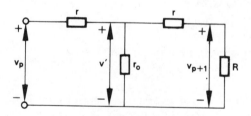

Fig. 3.16 Representative T-section with matching load attached

Evidently

$$\frac{v_p}{v'} = \frac{R}{R - r}.$$

By similar reasoning

$$\frac{v'}{v_{p+1}} = \frac{R + r}{R}.$$

Eliminating v' between these relations,

$$\frac{v_p}{v_{p+1}} = \frac{R + r}{R - r}.$$

If we now examine the circuit of Fig. 3.15, it can be seen that

$$\frac{v_0}{v_1} = \frac{v_1}{v_2} = \cdots = \frac{v_p}{v_{p+1}} = \cdots = \frac{v_{n-1}}{v_n} = \frac{R+r}{R-r}.$$

The product of these n voltage ratios is

$$\frac{v_0}{v_n} = \left(\frac{R+r}{R-r}\right)^n.$$

This is the voltage reduction ratio provided by the ladder, n being the number of T-sections in the ladder.

Because the resistance presented by the ladder to the source is equal to the resistance of the matching load, the ratio P_0/P_n of the powers delivered into these is equal to the ratio of the squares of the voltages developed across them. Thus

$$\frac{P_0}{P_n} = \left(\frac{v_0}{v_n}\right)^2 = \left(\frac{R+r}{R-r}\right)^{2n}.$$

This is the power reduction ratio. It is more meaningful to work in terms of the logarithm of this quantity:

$$\log_e\left(\frac{v_0}{v_n}\right)^2 = 2n \log_e\left(\frac{R+r}{R-r}\right) \text{nepers.}$$

More commonly the base 10 is used, in which case the power reduction ratio, or *attenuation* as it is called, is

$$\log_{10}\left(\frac{v_0}{v_n}\right)^2 = 2n \log_{10}\left(\frac{R+r}{R-r}\right)\text{bels} = 20n \log_{10}\left(\frac{R+r}{R-r}\right) \text{decibels.}$$

Notice the terms *nepers, bels* and *decibels* which are used here. The resistive ladder network is an example of an electrical attenuator.

3.7.3   Defects and limitations

Large changes in attenuation can be produced by suitable switching, which changes the effective number of T-sections without altering the input resistance. The matching requirement limits flexibility, and the manufacturer's calibration of an attenuator will be unreliable if an incorrect terminating load is used.

The characteristics of resistive ladders have been discussed here in terms of the response to dc sources, but the conclusions drawn are equally valid for ac and transient conditions, so that a wide range of applications is available. At increasingly high frequencies the performance is modified by stray inductance and capacitance in the resistors and switching circuitry.

In some alternative forms of resistive ladder networks there are modifications in the input and terminal circuits. The relative complexity of the ladder attenuator and difficulty in providing for small steps in attenuation cause alternative configurations such as the bridged-T network to be favoured in commercial designs.

Examples

1. Four resistors connected in series behave as 20 ohms. If the resistances of three of them are 7, 8 and 3 ohms, what is the resistance of the fourth?

2. Four conductances connected in parallel behave as 0.1 S. The conductances of three of them are 0.033, 0.027 and 0.020 S. What is the conductance of the fourth?

3. The electrical resistivity of copper is 0.2 microhm metres. What is the area of cross-section of a copper cable of length 10 kilometres and resistance 20 ohms?

4. What value of shunting resistance reduces 6 ohms effectively to 2 ohms? If a total current of 3 A flows in the combination, what current flows in each component?

5. What series resistance must be provided with a 200 ohm resistor, if the current drawn from a 300 V supply is to be limited to 1 A?

6. Show that for two resistances connected in parallel the current through either is given by the formula

$$\frac{\text{resistance of the other branch}}{\text{sum of the resistances}} \times (\text{total current}).$$

7. Two resistors are connected in parallel and carry a total current of 2.5 A. If one of them is of resistance 12 ohms and carries a current of 1.3 A, what is the resistance of the other?

8. Write down the battery currents for the networks of Fig. 3.17. All resistance values are in ohms. (It is quite easy to obtain the required answers by first recognising symmetries and other simplifying features.)

9. When a battery of emf 6 volts is connected across an 8 ohm resistor, the potential difference across its terminals falls to 4 volts. Calculate

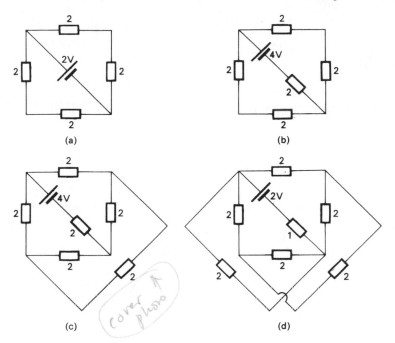

(a)   (b)

(c)   (d)

Fig. 3.17

the internal resistance of the battery. If the load is shunted by a resistance of 8/3 ohms, what is the new potential difference?

10. What is the emf of a dc dynamo which is designed to deliver maximum power when connected into a certain 12 V lighting system?

11. A secondary cell behaves as a source of emf of 2 V in series with a 20 ohm resistance. For what current does the potential difference across the terminals rise to 4 V, and what is the direction of flow?

12. The lamps in a 40 V lighting system consume a total current of 1.75 A. The system is supplied by a dynamo with internal resistance equal to the resistance presented by the load. What power is wasted in the dynamo?

13. Obtain a value for the current in each of the circuits of Fig. 3.18. All resistance values are in ohms.

14. Give values for the currents in each of the branches of the circuits of Fig. 3.19. All resistance values are in ohms.

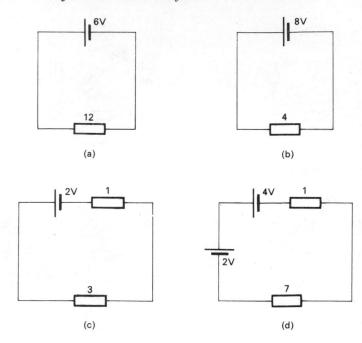

Fig. 3.18

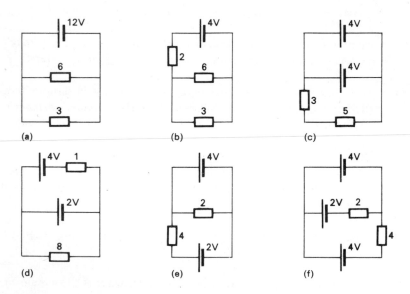

Fig. 3.19

15. Obtain values for the mesh currents in Fig. 3.20. What is the effect
     on these of removal of the 2 ohm resistor connected between *A* and
     B?

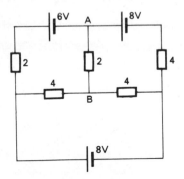

Fig. 3.20

16. In each circuit of Fig. 3.21, the load is the resistor at the right-hand
     end of the circuit. Identify the input resistance at the terminals, and
     calculate the voltage reduction (relative to the input) produced at
     the load. All resistance values are in ohms.

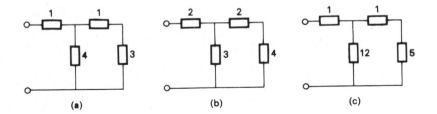

Fig. 3.21

17. If in the circuits of the previous question the input and load con-
     nections are interchanged, and the load is transferred to the other
     end of the ladder, is the input resistance altered?

18. For each of the circuits of Fig. 3.22, calculate
     (*a*) the input resistance,
     (*b*) the attenuation expressed as a voltage ratio, and
     (*c*) the attenuation in decibels.
     In each case the load is the resistor at the right-hand end, and all
     resistance values are in ohms.

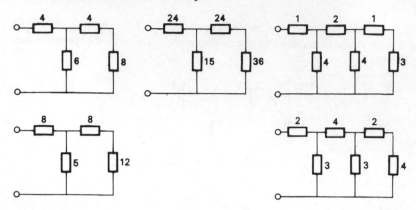

Fig. 3.22

19. Design a 40 Ω attenuator consisting of a single T-section which gives a power reduction ratio of 9 : 1 when correctly matched.

Steady current measurements

For most practical purposes an electric current is impermanent. One might consider preserving the ampere in terms of the calibrated scale of a moving-coil instrument, but even in the most favourable circumstances the unit so defined would be reliable to no better than one part in 10^4. In contrast, under optimum conditions the probable error associated with the emf of the Weston cell amounts to no more than two parts in 10^5, and the calibration of a good standard resistor is valid to two parts in 10^6. Thus, where high accuracy is sought under conditions of steady current, it is customary to measure voltage, current, resistance and power with the aid of a standard resistance or a standard cell, or both as may be appropriate.

4.1   Standard cell

The Weston cadmium cell was patented in 1892, and a present-day version is illustrated in Fig. 4.1. The positive and negative electrodes are respectively mercury and an amalgam of cadmium and mercury. The electrolyte is a solution of cadmium sulphate, which is separated from the mercury of the positive electrode by a depolarising layer of mercurous sulphate paste. In the saturated version of the cell, the strength of the electrolyte is maintained by cadmium sulphate crystals. The electrolyte is acidulated by the addition of sulphuric acid. The assembly and preparation of the cell follow a detailed specification employing very pure materials.

The emf varies slowly and predictably with temperature, which for precision measurements is maintained steady at 20 °C. The cell is not intended as a source of current. The internal resistance is usually several hundred ohms. It has been claimed that no permanent damage is caused by momentarily short-circuiting the electrodes, but the recovery period following the passage of excessive current is likely to exceed 24 hours. A limiting resistor of several thousand ohms should be connected in series with the cell until preliminary adjustments are complete, and the circuit

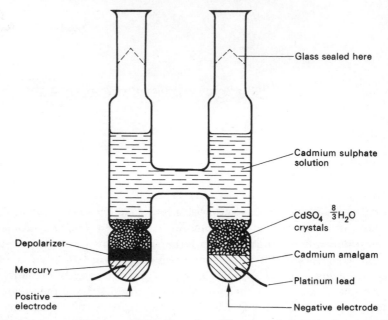

Fig. 4.1 Saturated Weston cell (Redrawn from Vigoureux, P. (1970), *Electrical Units and Standards*, HMSO.)

should be closed only momentarily when comparisons are made of emfs.

The useful life of the Weston cell exceeds ten years when it is correctly maintained, and can extend to several decades. The price of a commercially manufactured cell is only a few pounds, and it therefore constitutes a remarkably cheap investment.

4.2   Resistors

Resistance is by far the commonest circuit element, and is constructed in a considerable variety of forms. Except in power applications, the heat created by current-carrying resistors must be tolerated as an unavoidable by-product of their desired electrical function. The resulting local elevation of temperature may cause inconvenient changes in the values of electrical components.

4.2.1   Standard resistors

The calibration of a resistor intended as a precision standard should be performed in a good standards laboratory, and it is essential that the value then attached to it be permanent within an acceptably narrow range, and vary little with conditions of use (section 12.7.1). Standard resistors are

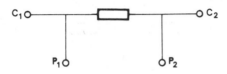

Fig. 4.2 Terminal connections for standard resistor

constructed with annealed wire of a copper alloy containing manganese and nickel. This has a low temperature coefficient of resistance, and exhibits great electrical stability. The thermal emf against copper is low, an essential feature in view of the prevalence of that metal in electric circuits.

Where the resistance value does not exceed about 10 ohms, separate terminals are provided for the current leads (C_1, C_2) and potential leads (P_1, P_2) (Fig. 4.2). A potentiometer draws zero current when correctly adjusted to measure a potential difference (section 4.4.1), so that any variation of contact resistance at the latter terminals is of no consequence.

Stability of resistance is likely to be better than a few parts in 10^7, provided the resistor is maintained with care and excessive currents are avoided in use.

4.2.2   Precision commercial resistors

High-grade wire-wound resistors are available commercially, which, while being inferior in specification to the standard resistors described above, will nevertheless meet the less exacting requirements appropriate for laboratory precision resistors and decade boxes. For the lower range of resistance values a cupro-nickel wire of medium resistivity is employed, while a nickel-chromium alloy is favoured for higher values. Both materials exhibit a very low temperature coefficient of resistance and a low thermal emf to copper, and are resistant to corrosion.

Wires are wound on bobbins or cards. The bobbin form of construction is compact. However, in situations involving time-dependence of current or voltage, stray capacitance must be minimised. Cards are very suitable for this purpose, and are therefore especially useful for pulse and alternating-current applications. The self inductance of card resistors can be reduced by doubling the wire back on itself before winding, or by reversing the sense of rotation of successive layers. Bobbin resistors are wound in sections to reduce self capacitance, the senses of rotation being alternately reversed to minimise self inductance. The range of resistance values normally extends up to only about 50 kΩ for card types, but is virtually unrestricted for bobbins. The nominal value of the resistance may be in error to the extent of five parts in 10^5, the long-term variation amounting to two parts in 10^5 per annum.

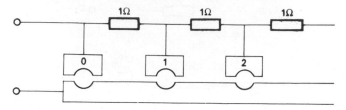

Fig. 4.3 Circuit of plug-type resistance box

4.2.3   Decade resistance boxes

In the simplest form of resistance box a number of resistors are connected as a series chain (Fig. 4.3). A selected combination is brought into use by inserting a brass plug in the appropriate socket. In high-quality boxes, precision resistors of the card type are used. These are contained within a metallic screening box connected to an additional terminal which can be earthed or connected to a chosen point in the circuit as desired. The combination of plug and socket is likely to offer appreciable contact resistance, so that a rotary switch is preferred except in carefully maintained precision apparatus. If there are eleven switch positions, the total resistance can be varied by steps of 0 to 10 units. As many as five decades are provided, so that in the absence of residual resistance the ratio of the maximum to the minimum non-zero resistance would be 10^5. This suggests that fifty accurate resistors might be needed. The number could be reduced by limiting the range to 0 to 9 units, but it would then be impossible to compare the calibrations of adjacent dials. The number of resistors per decade can be reduced from 10 to 6 without loss of the intercomparison facility, by suitably connecting to a *double-ganged* rotary switch. Values of residual resistances are often marked on resistance boxes by the manufacturer, and represent the resistance values presented at the terminals when the dials are set to zero.

For any setting of the decades, the maximum permissible current is limited by the highest resistance which is in circuit. Robust low-resistance switch-contacts are used, but some variable stray resistance inevitably remains, and there is little point in providing resistance steps smaller than 0.1 Ω.

The calibration of a good-quality resistance box may be reliable to five parts in 10^4 for high-resistance settings, but is significantly worse for total resistances below 10 ohms.

4.2.4   Radio-type resistors

Radio-type resistors are a large and varied group, designed for ease of mass-production and low price, with the additional advantage of minimum size consistent with intended wattage dissipation. Electrical stability is a secondary consideration, and is generally poor compared with standard and precision resistors. The use of radio-type resistors in measuring circuits should therefore be restricted to such functions as current limitation.

At the present time the resistive element is commonly composition, high-stability film, or wire-wound. This last is favoured for high-wattage applications. The wire is wound solenoidally on a cylindrical former of ceramic or glass, and is coated with an inorganic cement or vitreous enamel. Use is largely restricted to steady-current and power-frequency situations, as series inductance and shunt capacitance are inherently large.

Composite resistors are produced commercially in immense quantities and at low cost. The resistive element is a rod consisting of a mixture of carbon and various resins, and is suitably coated to prevent the ingress of atmospheric moisture which would otherwise temporarily raise the resistivity. The behaviour at high frequencies is quite good, and is better still for a film-type modification in which the composition coats the outer surface of a glass tube.

Metal and metal-oxide films are produced by high-temperature decomposition or vacuum deposition on to a ceramic base, followed by silicon lacquering. The same method of construction can also be used for the manufacture of precision resistors, as the electrical characteristics are very stable. For values of resistance below about 200 Ω, stray capacitance and inductance are relatively insignificant, and the electrical behaviour can be acceptable in applications involving frequencies ranging up to several hundred MHz.

4.2.5   *Variable resistors and potential dividers*

A resistance element provided with a slider can be used either as a variable resistor, or as a potential divider (Fig. 4.4). The former is essentially a

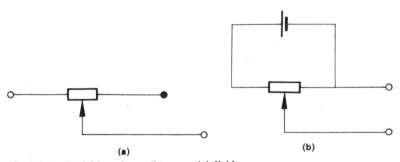

(a)   **(b)**

Fig. 4.4 *(a)* **Variable resistor,** *(b)* **potential divider**

two-terminal device, and the latter a three-terminal device with one terminal common to the input and output circuits. The potential divider is often called a potentiometer, a description which is unjustified in the absence of means of accurately comparing potential differences.

The term *rheostat* is used for a potential divider with the capacity to dissipate high powers, ranging perhaps from 10 to 1000 watts. Construction involves winding resistance wire on a ceramic core, and firing on a binding vitreous enamel. The current in the element will usually experience a discontinuity at the slider, so that in any application the limiting factor is current rather than total power.

For medium and low power uses the resistive element may be of the wire-wound, film, or solid composition type, and is usually mounted on a circular base, with the slider attached to a rotatable spindle. Non-wire-wound elements offer the advantage of low stray capacitance and inductance. Wire-wound elements give greater linearity and higher power-handling capacity, but the slider tends to move discontinuously from one wire to the next. Deliberate nonlinear variation with slider position can be achieved by providing varied geometry along the resistive track.

4.3   Instrument shunts and series resistors

Ammeters and voltmeters based on the moving-coil principle are essentially current-measuring devices, since the response is proportional to the current drawn. An instrument acquires the special characteristics appropriate to its intended role only when suitable external resistance is provided.

4.3.1   Voltmeters

A voltmeter measures the difference in potential between two points in a circuit, and when connected to these it shunts that part of the circuit which links them. The resulting change in the current distribution is accompanied by a fall in the voltage to be measured. If the readings obtained are to be meaningful, the instrument must therefore either remain connected throughout a series of measurements, or alternatively the resistance presented to the circuit must be made so high that the shunting effect is negligible.

The latter is the preferred technique, and voltmeters are commonly provided with a high resistance R connected in series with the moving coil (Fig. 4.5). For multi-range instruments a variety of series resistances can be selected by switching. These are sometimes called voltmeter multipliers.

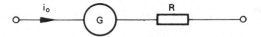

Fig. 4.5 Voltmeter circuit

The current i_0 which produces full-scale deflection is unchanged by the presence of series resistance, of course, although the effect of its addition is to raise the voltage corresponding to full-scale deflection to the value $i_0(R + G)$. Evidently stable resistors are essential if the instrument calibration is to be permanent.

Notice that the ratio

$$\frac{\text{resistance presented to circuit}}{\text{full-scale voltage reading}} = \frac{R + G}{i_0(R + G)} = \frac{1}{i_0}.$$

The current i_0 is small for a sensitive moving-coil instrument, and the reciprocal of i_0 is then large. The latter quantity is a useful figure of merit for the instrument, and is generally quoted in the units *ohms per volt*. Notice that the value is common to all ranges of a given multi-range voltmeter.

4.3.2 Ammeters

An ammeter must be connected *in series* with a circuit in order to determine the current which is flowing. The resistance of the coil can amount to as much as 20 Ω in an instrument designed to measure currents of a few milliamperes. This is of no great consequence in a circuit which is itself of high resistance, but will otherwise produce an appreciable reduction in the current to be measured. As with the voltmeter, the option then exists of connecting the instrument permanently in the circuit during a course of measurements.

Where the sensitivity greatly exceeds requirement, it is usual to provide a series resistance R and a relatively low shunting resistance r (Fig. 4.6). If several shunts are supplied there is the additional advantage of variety of ranges. For stability of calibration the resistance R should be large compared with G, and both R and r should possess low temperature

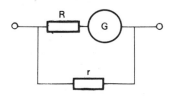

Fig. 4.6 Shunted moving-coil instrument

coefficients and be generally stable electrically. The function of R is to reduce the effects of any changes in G. Favoured alternatives for the material of R and r are manganin and a manganin alloy. For high current ranges the shunt may take the form of a flat strip, which provides a large heat-radiating surface area.

4.3.3   Universal shunt

When the coil of a moving-coil instrument rotates, an emf is induced in it which drives current through the attached circuit. The current causes dissipation of energy in the circuit as heat. This takes place at the expense of the rotational energy of the coil, and therefore 'damps' its motion. Damping can be turned to advantage by using it to reduce the waiting time while the coil comes to rest. Moving-coil instruments are often provided with built-in damping by winding the coil on a conducting former. Alternatively a suitable value of resistance is connected across the terminals. The damping effect increases as the resistance is reduced, but the time taken for the coil to come to rest passes through a minimum for a particular value of resistance, and the system is then said to be critically damped (section 5.4.5). A slightly under-damped condition is usually favoured.

For a multi-range instrument, the value of the shunt resistance would normally be altered by a change of range, unless there is simultaneous adjustment of a resistance provided in series with the moving coil. This additional feature appears in the Ayrton universal shunt of Fig. 4.7(*b*). Consider first the simpler arrangement of Fig. 4.7(*a*). Let i_0 be the current which produces full-scale deflection of the coil, and suppose the actual current drawn by the instrument is then i. The potential difference across the terminals is

$$i_0(G + y) = (i - i_0)x.$$

If $x + y$ is chosen to be equal to the critical damping resistance S, then

$$i_0(G + S) = ix.$$

If now the range of the instrument can be changed without alteration of S, the right-hand side of this equation remains constant. Thus for a sequence of ranges of current bearing the constant ratio $10:1$, the values of x fall progressively in the ratio $1:10$. This conclusion in unaffected by the parameters of the moving-coil instrument used, and in this sense the shunt is 'universal'. A given shunt is however suitable only for an instrument for which it provides the correct damping resistance.

The shunt of Fig. 4.7(*b*) gives an external damping resistance of $1\ \mathrm{k}\Omega$, and range ratios of $10:1$. Damping will of course be increased if the

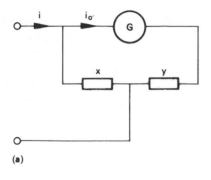

(a)

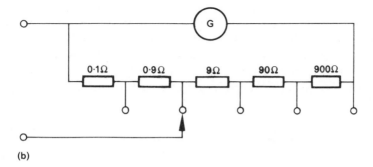

(b)

Fig. 4.7 **Ammeter provided with** *(a)* **single divided shunt,** *(b)* **Ayrton shunt**

circuit from which the current is derived contributes appreciable additional shunting action.

4.3.4 *Multi-test instruments*

Robust multi-test instruments based on the moving-coil principle are widely used. Shunts and multipliers are selected by switching, enabling a wide range of currents and voltages to be measured. Some instruments give full-scale deflection for currents as low as 10 μA, and voltage ranges may present internal resistances as high as 100 kΩ per volt. With an internal battery switched into circuit, resistance can be measured by connecting it across the terminals of the instrument and noting the current which passes, as recorded on a scale calibrated in terms of resistance. An internal rectifier provides conversion from dc to ac operation, although the accuracy of calibration deteriorates at frequencies in excess of a few kHz.

4.4   DC potentiometers

A potentiometer is an instrument used for measuring an unknown emf or potential difference by balancing it against the known potential difference produced by a flow of current in a network of known characteristics. This definition applies equally well to dc and ac potentiometers, although we shall restrict ourselves here to the former. The incorrect use of the term 'potentiometer' for three-terminal radio resistors has already been noted (section 4.2.5).

4.4.1   Simple potentiometer

The essential elements of a dc potentiometer are illustrated in Fig. 4.8. A steady current derived from a driver cell of emf v_0 flows through a uniform resistance wire AB. The potential difference developed across the wire can be adjusted by means of the rheostat R. A Weston cell of emf v_s is provided, and the emf v of the third cell appearing in the diagram represents the unknown emf or potential difference which is to be measured. Notice that electrodes of similar polarity are connected to the end A of the wire. G is a sensitive moving-coil centre-zero galvanometer.

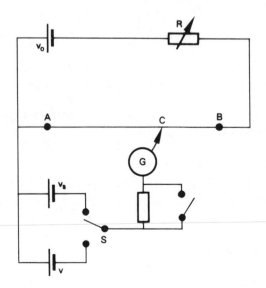

Fig. 4.8 Basic potentiometer circuit

If the standard cell is switched into circuit at S, and the position of the slider is adjusted so that the galvanometer gives a null response, then the potential difference across the length l_s of wire between the slider and end A is equal to v_s. This part of the procedure is called *standardisation* because it gives the voltage drop per unit length of wire in absolute terms.

Suppose that when the unknown is switched into circuit in place of the standard cell a new balance length l is obtained. Then

$$\frac{v}{v_s} = \frac{l}{l_s},$$

or

$$v = l\,\frac{v_s}{l_s}.$$

Since no current passes through the galvanometer at balance, the method is independent of contact resistance at the slider and of the sensitivity of the galvanometer, except in so far as the latter may limit the accuracy with which the correct balance length can be identified. Null response is a condition which is employed quite generally in bridge measurements. In fact the term 'bridge' refers to the 'bridging' of two terminals by the detector, with a view to determining whether a difference in potential exists between them (section 10.1). The wire of the potentiometer is usually called the bridge wire.

An additional advantage deriving from the dependence on a condition of null response is that no current is then being drawn from the standard cell or the unknown. For the former it is the actual emf which is measured, regardless of any internal series resistance which it may possess. It is essential that minimum current be drawn from the standard cell at all times, and a protective resistance should be provided in series with the galvanometer, to be shorted out only when the bridge is close to balance. The sensitivity of the galvanometer itself can be conveniently controlled with a universal shunt.

The method enables the unknown voltage to be related directly to the emf of the standard cell. The accuracy depends on the uniformity of the wire, and the absence of stray resistance at end A. Sources of error include uncertainty in the position of the slider for balance, and in the precise location of the end A of the wire. These last errors can be reduced by using a very long bridge wire. It is important that the driver cell should have the capacity to provide a constant current over the period of time occupied by a set of measurements, and the standardisation should be checked frequently with the standard cell so that there is warning of drift. An electronically stabilised driver source could be advantageous.

The bridge wire could be replaced by two decade resistance boxes connected in series, balance being obtained by adjusting each in such a way that their sum remains constant. Some of the disadvantages mentioned in the previous paragraph can be eliminated in this way, but overall accuracy now depends on the separate calibration accuracies and residual resistances of the two boxes.

The accuracy of measurement of small emfs is additionally limited by the presence of thermal emfs, which may amount to a few microvolts. Metallic contacts and conductors should be carefully matched with a view to minimising this source of error. Accuracy can be further improved by repeating each measurement with the polarities of the driver cell and unknown reversed, and averaging.

4.4.2   Crompton potentiometer

The number of significant figures obtainable in potentiometer measurements can be increased by lengthening the bridge wire. The same effect is obtained by the addition of a fixed series resistor, which for obvious reasons is sometimes called a 'wire-stretcher'. In a modification suggested by Crompton in 1893, several resistors are provided (Fig. 4.9), each equal in resistance to the slide wire. The arrangement can be made direct-reading by connecting the Weston cell in circuit with one terminal attached to the galvanometer and the other to the stud marked 1.0. Move the slider 0.186 of the length of the bridge wire away from end A. The circuit is then balanced by adjustment of the rheostat. If the emf of the standard cell is 1.0186 V, and the bridge wire is 1 metre long, the voltage drop per unit length on the wire is 0.1 mV per mm, and voltages up to 1.5 V can be measured.

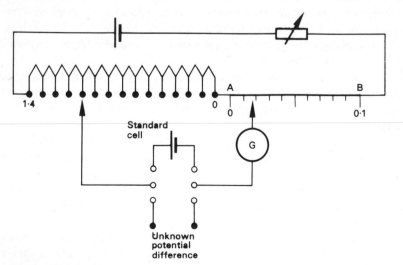

Fig. 4.9 Crompton potentiometer

The studs attached to the fixed resistors carry no current at balance, but the rheostat must be designed for minimum variability of contact resistance. Errors caused by changes in temperature are compensated by constructing all resistors with manganin wire, although a hard-wearing platinum-silver alloy is preferred for the slide-wire.

4.4.3   Kelvin—Varley potential divider

The Kelvin—Varley potential divider is a resistance network which permits extremely fine subdivision of potential difference while presenting constant resistance to the driver cell. Each of the four decades of the example illustrated (Fig. 4.10) is composed of eleven equal resistors, and each decade is connected in parallel with two resistors of the decade above it. The combination of the 1.6 Ω slide wire in parallel with two of the 0.8 Ω resistors of the decade above produces an effective resistance of 0.8 Ω, so that the decade presents a total resistance of $10 \times 0.8\ \Omega = 8\ \Omega$. Following through the same line of reasoning with successive decades, it can be seen that each behaves as ten equal series resistors, instead of the eleven which are actually present.

If the potential difference supplied to the top decade is 1 V, then the potential differences developed across successive decades taken in order from top to bottom are

$$1\ \text{V},\ 0.1\ \text{V},\ 0.01\ \text{V}\quad\text{and}\quad 0.001\ \text{V},$$

and the potential difference developed across the slide wire is 10^{-4} V. The decade switches enable readings to be made to four significant figures, and additional figures are obtained with the aid of the slide wire.

The number of decades which can usefully be provided is limited by stray resistance at the stud contacts.

The network is standardised by connecting a Weston cell in place of the source of unknown potential difference and adjusting the rheostat for zero galvanometer deflection, with the dials set to read the known emf of the cell.

4.4.4   A modern potentiometer

The standard cell in the potentiometer of Fig. 4.11 is permanently connected. Any drift of the attached centre-zero galvanometer can be immediately corrected by adjustment of the rheostat in the driver unit, so that standardisation of the measuring circuit is maintained continuously in terms of the emf of the standard cell and the known values of the resistances in the standardising circuit. An additional variable resistor can be

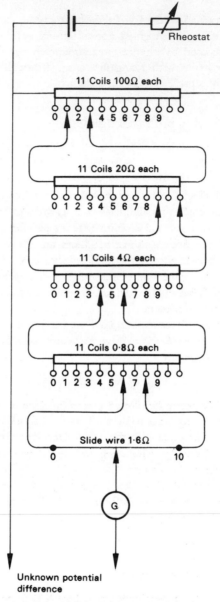

Fig. 4.10 Four-decade Kelvin—Varley divider with slide-wire

provided in the standardising circuit, with which adjustment can be made to compensate for the temperature-dependence of the emf of the standard cell.

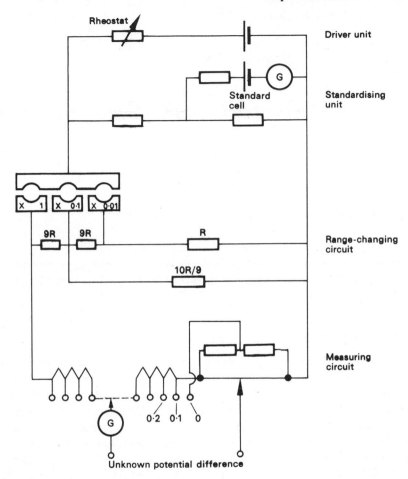

Fig. 4.11 Modern potentiometer

A true zero is not usually available on the slide-wire of a potentiometer, because of contact resistance at the zero end. In the potentiometer illustrated a shunt constructed of two series resistors is connected in parallel with the slide-wire, and a lead is brought out from their common point and connected to a stud identified as 'zero'. The slider can then provide a small negative reading, as well as a range of positive readings. The resistance of the slide-wire with its shunt is chosen to be slightly greater than that of each step on the decade.

The resistance presented by the measuring circuit at balance is R. It is quite easy to show that as the plug is transferred successively from the X 1 to the X 0.1 and X 0.01 positions, then the current in the measuring

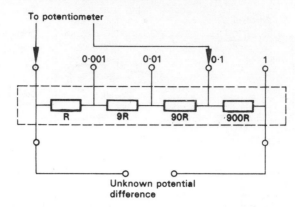

Fig. 4.12 Use of volt box for division of potential difference

circuit is reduced to 0.1 and 0.01 of its original value. The *combined* current drawn by the measuring- and range-changing circuits is meanwhile unchanged. The voltage range is in consequence reduced in the ratio 1/10 and 1/100.

4.4.5 Potentiometer measurements

The most direct application of the potentiometer is the measurement of voltage, and the instrument is used extensively for this purpose in a considerable variety of fields of electrical measurement, including the determination of thermal and voltaic emfs, and the calibration of voltmeters.

Large voltages must be scaled down before being presented to the potentiometer, and for this purpose a voltage-ratio box is used. This is known more commonly as a volt box, and provides a number of accurately calibrated series-connected resistors (Fig. 4.12), which can be connected across the unknown source. In the example illustrated, fractions 0.1, 0.01, etc., of the unknown voltage can be presented to the potentiometer. The latter draws no current at balance, so that the division of voltage depends accurately on a known resistance ratio. Current is drawn from the un-

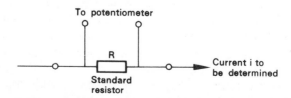

Fig. 4.13 Measurement of current

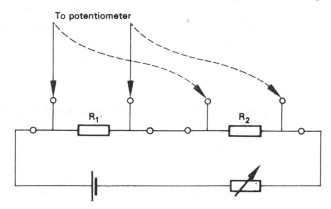

Fig. 4.14 Circuit for comparison of resistances

known regardless of whether the potentiometer is balanced, so that the technique is not a null one, and R should preferably be large.

Current i can be measured using a standardised potentiometer to determine the potential difference iR which it produces across a four-terminal standard resistor R (Fig. 4.13). An ammeter can be calibrated with the same arrangement by connecting it in series with the source of current, which is then varied over the desired range.

Two resistances can be compared with each other, using the circuit of Fig. 4.14. Their ratio is equal to the ratio of the potential differences developed across each, and this is determined with a potentiometer. The resistances should preferably be of a similar order of magnitude and of the four-terminal type, and if one is a standard resistance, then the method standardises the other.

4.5   Wheatstone bridge

The Wheatstone bridge was devised by Christie in 1833, but remained virtually unknown until brought into prominence ten years later by Sir Charles Wheatstone. The bridge is widely used for the comparison of resistances. It is a logical development of the potentiometer, and is especially suited for the determination of medium resistance values. Special versions enable resistance to be measured over a range extending from 1 Ω to more than 1 MΩ, the accuracy under ideal conditions being better than one part in 10^7.

The bridge is the parent of a number of dc and ac bridges, all using basically a closed loop comprising four arms, with a source and detector inserted in opposite diagonals.

4.5.1   Balance condition

The basic circuit is that of Fig. 4.15. If P and Q are the resistances to be compared, then R and S are called the ratio arms. The battery could be a 2 V accumulator or 1.5 V dry cell. G is a sensitive moving-coil galvanometer.

   If no current flows in the galvanometer the bridge is said to be balanced, and the following simple relations hold for the currents indicated in the diagram:

$$i_P = i_Q, \quad \text{and} \quad i_S = i_R.$$

Since there is then no difference in potential between the two ends of the galvanometer branch, it follows that

$$Pi_P = Si_S, \quad \text{and} \quad Qi_Q = Ri_R.$$

Elimination of the currents between the above four equations gives

$$\frac{P}{Q} = \frac{S}{R}.$$

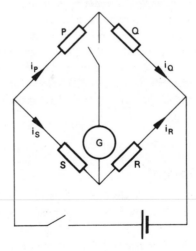

Fig. 4.15 Basic Wheatstone bridge

This is the balance condition. It is independent of the characteristics of the galvanometer and battery, and is therefore unaffected by the magnitude and stability of the emf of the latter. It is easy to show that the same balance condition is obtained if the positions of the battery and galvanometer are interchanged.

4.5.2   Balancing procedure

The balancing procedure involves adjustment of one or more of the resistors in the bridge. In commercial patterns several values can be selected for R and S by switching, ranging upwards from one ohm by factors of ten. If P is an unknown resistance, then Q is conveniently a resistance comprising four or five decades.

Separate keys are often provided for the battery and the galvanometer. The battery key should be closed first, so that transient effects associated with stray inductance die away before the galvanometer is connected into the circuit (section 5.1). The galvanometer key should in any case be closed only briefly, to minimise the risk of damage caused by excessive current. As an extra precaution a universal shunt or series resistance should be connected to the instrument until the bridge is very nearly balanced. A shorting key is connected across the galvanometer so that the instrument can be brought rapidly to rest.

Errors due to thermal emfs can be eliminated by repeating the balancing procedure with the battery leads interchanged, and averaging the two results obtained. Precision commercial bridges are calibrated at 20 °C, and when used at this temperature can provide accuracy better than 0.01%.

The following procedure is suggested for the measurement of an unknown resistance:

(*a*) First obtain a rough balance by suitable adjustment of the bridge arms. Calculate an approximate value for the unknown resistance P.

(*b*) The galvanometer responds most sensitively to a small change in the resistance of any arm when the resistances of the four arms are of the same order of magnitude. This is a rough rule which suffices for the limited purpose here. Set the three adjustable arms to be approximately equal to P. It will now be found that a balance can be selected quite critically by adjustment of Q. By way of example, suppose that at balance: $R = S = 100\ \Omega$, and $Q = 101\ \Omega$. Then $P = 101\ \Omega$.

(*c*) If no subdivided ohm is available, an additional significant figure can be obtained by making $R = 1000\ \Omega$, and again adjusting Q for balance. Suppose that $S = 100\ \Omega$ and $Q = 1009\ \Omega$. Then $P = 100.9\ \Omega$.

(*d*) The number of significant figures can be extended by further reduction of the ratio S/R, but a limit is eventually set by the accompanying fall in sensitivity.

4.5.3   Sensitivity

A bridge is said to be sensitive when a relatively large off-balance current is produced by a given small fractional change in the resistance of one arm of

the bridge. The greater the sensitivity, the greater the number of significant figures which can be determined for the value of the resistance which
is being measured. An additional factor contributing to the overall sensitivity is the *current sensitivity* of the galvanometer.

   If sensitivity were all-important, the maximum power theorem
would require (section 3.6) that the resistance presented by the bridge to
the battery should equal the internal resistance of the latter. But some dc
sources would be destroyed by the excessive current which would flow if
they were used in this way, and there could also be damage to resistance
coils. On this latter account the magnitude of the source emf should in any
case always be carefully limited.

   The galvanometer resistance usually exceeds the internal resistance
of the battery, and greater sensitivity is then obtained by connecting one
terminal of the galvanometer to the junction of the two largest resistances,
and the other (necessarily) to the junction of the two least resistances.
This is a special case of a theorem which is valid for ac and dc bridges
generally.

4.6   Carey—Foster bridge

The Carey—Foster bridge (Fig. 4.16) is a slide-wire modification of the
Wheatstone bridge, with which the differences between two nearly equal
resistances R and S can be accurately determined. It is especially useful for
the calibration of a resistor in terms of a standard of about 1 ohm.

   The bridge is balanced by adjustment of the position of the slider,

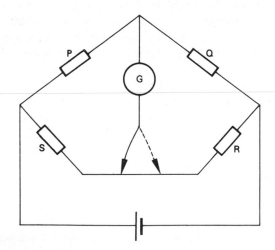

Fig. 4.16 Carey—Foster bridge

with the components connected as indicated. Now let R and S be interchanged, and suppose the slider has to be moved a distance l to the right to rebalance the bridge. The interchange has caused an amount of resistance $S-R$ to be transferred to the right-hand side of the slide-wire. If the resistance per unit length of the latter is r, this has been compensated by moving an amount of wire of resistance lr to the left-hand side of the slider. Therefore

$$S-R = lr.$$

The necessary circuit alterations should be made either by switching, or by interchanging the actual positions of R and S, and it is important that lead resistances should not be transferred.

A value can be obtained for r either by connecting a small known resistance successively in series with R and S, or by shunting each successively with a large known resistance. The change in balance position is related as a length of bridge wire to what is effectively a known resistance. The calibration can be extended along the length of the slide-wire by repeating the process with various values for the ratio P/Q.

A sensitive galvanometer is required, and the usual precautions should be taken to minimise and compensate for the effects of thermal emfs.

4.7   Kelvin double bridge

The Kelvin double bridge is suitable for the comparison of four-terminal resistances in the range 0.1 to 0.001 Ω. The circuit is that of Fig. 4.17.

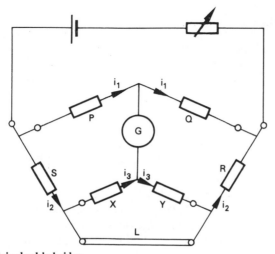

Fig. 4.17 Kelvin double bridge

The low resistances R and S are connected by the thick copper link L. The battery drives a current of several amperes through the combination, the actual magnitude being controlled by the rheostat. The current leads of R and S carry the heavy current, and current is also drawn through the potential leads. The resistances P, Q, X and Y are large compared with any contact or lead resistances which may be present.

Several alternative methods are available for balancing the bridge. In one of these the resistances P and X are adjusted until the bridge is balanced whether the link L is out or in. In the absence of the link the circuit is an ordinary Wheatstone bridge, in which stray resistances can be neglected in comparison with the larger resistances which are present in each arm. The condition of balance is

$$\frac{P}{Q} = \frac{S + X}{R + Y}.$$

When the link is present the following relations hold:

$$Pi_1 = Si_2 + Xi_3, \quad \text{and} \quad Qi_1 = Ri_2 + Yi_3.$$

It follows from the above three equations that

$$\frac{P}{Q} = \frac{S + X}{R + Y} = \frac{S + Xi_3/i_2}{R + Yi_3/i_2}.$$

These relations are satisfied provided

$$\frac{P}{Q} = \frac{S}{R} = \frac{X}{Y},$$

from which the ratio $\dfrac{S}{R}$ can be determined.

Additional reading

BALDWIN, C. T., *Fundamentals of Electrical Measurements.* Harrap, 1961.

BUCKINGHAM, H., and PRICE, E. M., *Principles of Electrical Measurements.* English Universities Press, 1967.

DUFFIN, W. J., *Electricity and Magnetism.* McGraw-Hill, 1965.

GALL, D. C., *Direct and Alternating Current Potentiometer Measurements.* Chapman and Hall, 1938.

GOLDING, E. W., and WIDDIS, F. C., *Electrical Measurements and Measuring Instruments.* Pitman, 1963.

HAGUE, B., and FOORD, T. R., *Alternating Current Bridge Methods.* Pitman, 1971.

HARNWELL, G. P., *Principles of Electricity and Electromagnetism.* McGraw-Hill, 1949.

HARRIS, F. K., *Electrical Measurements.* Wiley, 1962.

KARO, D., *Electrical Measurements.* Part II. Macdonald, 1953.

SHIRE, E. S., *Classical Electricity and Magnetism.* Cambridge University Press, 1960.

THOMAS, H. E., *Handbook for Electronic Engineers and Technicians.* Prentice-Hall, 1965.

THOMPSON, J. R., *Precision Electrical Measurements in Industry.* Butterworths, 1965.

TURNER, R. P., *Bridges and other Null Devices.* Foulsham-Sams, 1968.

VIGOUREUX, P., *Electric Units and Standards.* HMSO 1970 (National Physical Laboratory).

VIGOUREUX, P., *Units and Standards for Electromagnetism.* Wykeham Publications, 1971.

WELLARD, C. L., *Resistance and Resistors.* McGraw-Hill, 1960.

SPECIAL PUBLICATION, NBS No. 300, Vol. 3. HERMACH, F. L. and DZIUBA, R. F. (eds.) *Precision Measurement and Calibration. Electricity − Low Frequency,* 1969.

Examples

1. A potential divider consists of a 200 ohm and a 300 ohm resistor connected in series across a 12 volt supply. What current flows in a 600 ohm load connected across the 300 ohm resistor?

2. The current through a load resistor r is drawn from a source of constant voltage v_0, and is controlled by a series resistor adjustable in value from zero to R. What is the available range of potential difference across r?

   Alternatively, the resistor R is connected across the supply, with a contact which can slide along it from one end to the other. r is connected to the slider and to one end of R. If $r = R = 100 \ \Omega$ and $v_0 = 1$ V, what maximum current must R be able to carry?

3. A 300 ohm potential divider is connected across a 6 V supply. What is the standing current in the absence of any additional load? Where is the appropriate tapping point for a high resistance load across which a potential difference of 1 V is required? If alternatively a load of 200 ohms is provided, where is the appropriate tapping point for a potential difference of 3 V? In this case what currents flow in the load and from the source?

4. A 300 ohm load is to be supplied with a variable current drawn from a 32 V dc supply of negligible internal resistance. Give diagrams showing how this may be arranged with a rheostat connected
    (*a*) as a potential divider, and
    (*b*) as a variable series resistance.
    Suggest a disadvantage of each circuit.
        If in case (*a*) a 125 ohm rheostat is used, what total current is drawn from the supply when the load current is 80 mA? In case (*b*) what is the power drawn from the supply when the dissipation in the load is 0.48 W?

5. The resistance of an ammeter reading 2A FSD is 0.1 Ω. What maximum potential difference develops across the instrument?

6. The resistance of a voltmeter is 1 kΩ per volt. If it reads 100 V FSD, what is its resistance, and what maximum current does it draw?

7. The resistance of the moving coil of a voltmeter is 5 Ω, and a series resistance of 995 Ω is provided. If the full scale reading is 1 V, what is the maximum current in the coil? How could the instrument be simply modified to read 10 V FSD?

8. Explain how a 10 Ω moving-coil instrument which reads 50 mA FSD can be simply modified to read
    (*a*) 1 V FSD,
    (*b*) 10 V FSD, and
    (*c*) 100 V FSD.

9. The resistance of the moving coil of an ammeter is 6 Ω, and it reads 50 μA FSD. How could it be simply modified to read 150 μA FSD? What is then the effective resistance of the instrument?

10. How could you most simply modify an instrument reading 50 mA FSD and having moving-coil resistance 10 Ω to read
    (*a*) 100 mA FSD,
    (*b*) 550 mA FSD, and
    (*c*) 1.05 A FSD?

11. A moving-coil instrument has a coil resistance of 5 Ω, and reads 5 mA FSD. It is critically damped by an external circuit resistance of 15 Ω. Give a single divided shunt design enabling the instrument to read 10 mA, 100 mA, 1 A and 10 A FSD.

12. If the arms of a Wheatstone bridge taken in cyclic order are of resistance 64, 64, 100 and 100 ohms, and the galvanometer is connected where equal bridge arms meet, what galvanometer resistance would match the bridge?

Transients

5.1 Introduction

If it were possible to make up a circuit in the form of a single loop containing only pure sources of emf and pure ohmic resistors, then as soon as the last connection was completed the current would change abruptly from zero, and remain steady at the new value indefinitely. In practice resistance values alter with temperature, and batteries run down, so that if an established current is to be held steady some kind of electrical compensation must be provided.

But there are additional factors which make it difficult even to produce the initial abrupt change in current, and it is with these that we shall be specially concerned in this chapter. They may be inherent in the construction of the circuit components, or in the layout of the connecting leads. Their effect is the prevention of sudden change, so that the initial and final end states of a circuit are linked by a relatively gradual *transient response* as it is called, rather than by an abrupt step.

In dc electrical bridges, one is advised to depress the battery key a little before the detector key (section 4.5.2), so that transient effects are virtually extinct before the detector is in circuit. Such transient phenomena are generally of quite short duration, ranging commonly from a fraction of a millisecond up to a few seconds.

The form of a transient response may be alternating or unidirectional, depending on the character of the circuit and the nature of the excitation. The effective duration before steady-state behaviour becomes predominant is dependent on the constants of the circuit. In this context the term *steady state* is used for situations involving either dc steady currents or alternating currents of constant amplitude.

At one extreme there are transient responses which can be destructively harmful, such as switching surges in power systems, while at the other there are innumerable useful applications. Electronic generators for all kinds of wave-shapes rely largely on the exploitation of transient effects.

Transient phenomena depend for their existence on the presence in a circuit of capacitance and/or inductance. These are properties whose exist-ence we have not found it necessary to take into account in earlier chapters, but which are always present, either by deliberate introduction or as accidental strays. We shall turn our attention first to capacitance.

5.2   Capacitance

A simple capacitor can be visualised as a pair of parallel conducting plates, fixed preferably quite close together, and separated by an insulating medium such as air or vacuum. A conducting wire attached to each plate enables connections to be made to an electric circuit.

Suppose the device is connected into the circuit of Fig. 5.1, so that it forms a closed loop with a source of emf v_0, a resistor R, and a key.

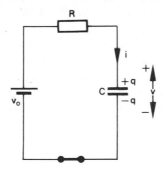

Fig. 5.1   Charging of a capacitor

When the key is closed, an electric current flows for a while, in the sense indicated. As the capacitor is electrically an open circuit, when current flow ceases the difference in potential between the plates must equal the emf of the battery.

The temporary flow of current could be detected by placing a moving-coil instrument in series with the circuit. The deflection would be always in one direction, indicating that the flow itself is unidirectional, and that therefore a net flow of charge takes place round the circuit. The sense of current flow is such as to produce an accumulation of positive charge on the plate marked positive, and of negative charge of equal magnitude on the plate marked negative. In effect the process results in a *gain* of positive charge by the positive plate and a *loss* by the other, and it is quite usual in conversation to refer unspecifically to the charge on a capacitor.

The charge can be measured by discharging the capacitor through a calibrated ballistic galvanometer. The moving coil of this instrument registers a throw proportional to charge. For a given capacitor, the stored charge q is found to be proportional to the applied potential difference v, so that we may conveniently write

$$q = Cv,$$

where C is a constant of the capacitor, and is known as the capacitance. A capacitor is sometimes called a condenser, which is its older name (section 1.4).

The value of C can be made large by the use of (*a*) large plate areas, (*b*) small clearance between plates, and (*c*) the introduction of suitable insulating material, known as dielectric, between the plates.

We have seen that the units of charge are coulombs, if the units of current and time are amperes and seconds respectively. If the potential difference v is in volts, then the units of C are farads. One farad may therefore be defined as the capacitance of a condenser across which a potential difference of one volt exists when the charge stored is one coulomb. The farad is such a large unit that in a lifetime of work with electrical apparatus one might never encounter a capacitance of such immense magnitude. Capacitance values are usually quoted in terms of the microfarad (μF, 10^{-6} F), nanofarad (nF, 10^{-9} F), or picofarad (pF, 10^{-12} F).

5.2.1   Stored energy

When a capacitor is charged by a battery, current flows for a time and the battery loses energy. Some of this energy is dissipated as joule heat in ohmic resistance in the circuit, and some is stored by the capacitor. If the potential difference is large enough, a spark passes when the capacitor is short-circuited after disconnection from the circuit. A large capacitor might even briefly light a lamp. The heavy current associated with rapid discharging can cause mechanical damage in capacitors. In comparison with chemical cells, the energy which can be stored in a given volume in this way is very small.

During the process of charging the capacitor of Fig. 5.2(*a*), the instantaneous current is

$$i = \frac{\mathrm{d}q}{\mathrm{d}t}.$$

This is effectively Kirchhoff's first law for the connection between the circuit and the positive plate. Also

$$q = Cv,$$

so that

$$i = C \frac{dv}{dt}.$$

The conventions used in the diagram for the direction of the current and the polarity of the potential difference should be compared with those of Fig. 5.2(*b*) for a resistor, for which the voltage–current relation is of course

$$v = iR.$$

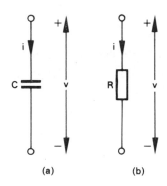

(a)   (b)

Fig. 5.2 Current flow in (*a*) **capacitor,** (*b*) **resistor**

The rate of supply of energy to the capacitor is

$$P = vi = vC \frac{dv}{dt},$$

and the energy supplied in time d*t* is

$$dW = P \, dt = vC \, dv.$$

The total energy stored at potential difference *v* is

$$W = \int_0^v vC \, dv = \tfrac{1}{2}Cv^2 = \tfrac{1}{2}qv = \tfrac{1}{2}\frac{q^2}{C},$$

where the relation $q = Cv$ has been used to obtain the alternative expressions. The units of *W* in the above are of course joules.

When the steady state is reached, the potential difference across the capacitor is equal to the battery emf v_0, so that the energy stored is $\frac{1}{2} C v_0^2$ and is independent of any resistance in the circuit. On the other hand the battery has supplied charge q at constant potential difference v_0, so that it has delivered energy $q v_0 = C v_0^2$ into the circuit. An amount of energy $\frac{1}{2} C v_0^2$ has still to be accounted for.

The above formulae indicate that whatever the rate of charging, a condenser can take up only half the energy supplied by a source of constant emf. The lost energy is usually mostly consumed by circuit resistance. In the extreme case of a circuit of negligible resistance, the initial charging current is large and of brief duration. The circuit acts as an aerial, the wasted energy being partly radiated away as a pulse of electromagnetic radiation. The spark which may form at the switch contacts as the circuit is closed consumes some of the energy in the form of light and heat.

5.2.2   Capacitors in parallel

In Fig. 5.3(*a*) two capacitors are shown connected in parallel. The applied potential difference v is common to both. The total charge q is the sum of the charges on each, so that

$$q = q_1 + q_2.$$

Let C be the effective capacitance of the combination. Then

$$q = Cv = C_1 v + C_2 v,$$

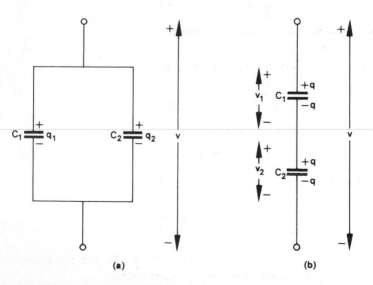

(a)   (b)

Fig. 5.3 Capacitors connected (*a*) in parallel, (*b*) in series

giving

$$C = C_1 + C_2.$$

It is easy to extend this result for any number of capacitors $C_1, C_2, \ldots$ $C_k, \ldots, C_n$ connected in parallel, for which the effective capacitance is

$$C = \sum_1^n C_k.$$

5.2.3 Capacitors in series

Consider the arrangement of Fig. 5.3(b) in which two capacitors are shown connected in series. Suppose for the moment that no source of emf is present and that the capacitors are initially uncharged. When a source of emf is connected, the total charge on the two capacitor plates which are directly connected together will still be zero, for these are insulated from the rest of the circuit. The separate charges which they carry must therefore be equal in magnitude and opposite in sign. This is a consequence of charge conservation (section 2.1.5). Moreover, the plates of either capacitor also carry equal and opposite charges, so that the charge distribution must be as shown in the diagram. Notice that the sum of the charges on the two plates connected directly to the battery is also zero, as one would expect.

It appears therefore that the individual capacitors in a series combination each carry the same charge. The applied difference in potential v is distributed as two steps in potential v_1 and v_2, so that

$$v = v_1 + v_2.$$

If the effective capacitance of the combination is C, this relation can be written

$$\frac{q}{C} = \frac{q}{C_1} + \frac{q}{C_2},$$

or

$$\frac{1}{C} = \frac{1}{C_1} + \frac{1}{C_2}.$$

It is easy to extend this result to the general case of capacitors $C_1, C_2, \ldots$ $C_k, \ldots, C_n$, connected in series, for which the effective capacitance is given by the relation

$$\frac{1}{C} = \sum_1^n \frac{1}{C_k}.$$

The electrical insulation is never perfect in a capacitor, so that a

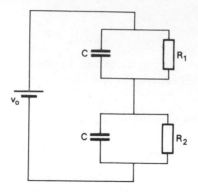

Fig. 5.4 Series combination of capacitors, with leakage resistors

small current is likely to be conducted between the plates whenever a difference in potential is established. This can distort quite markedly the potential and charge distributions in a series combination. Consider the arrangement of Fig. 5.4. This represents a series combination of two capacitors which have equal capacitances C and differing shunt leakage resistances R_1 and R_2. When a battery is connected across the combination it is the resistors which determine the equilibrium potential distribution, for these will go on carrying current indefinitely. The potential differences across R_1 and R_2 are easily shown in consequence to be $v_0 R_1/(R_1 + R_2)$ and $v_0 R_2/(R_1 + R_2)$ respectively when steady conditions have been attained. The capacitors, although charged, cannot modify this situation, since in the steady state each behaves as an open circuit.

These conclusions become important when a combination of series capacitors is operated at voltages which approach the breakdown point of any of the individual components. If the leakage conductance of one of the capacitors is large, then other components in the chain may be subjected to excessive voltages. One method of correction consists in con-

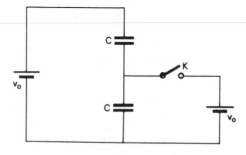

Fig. 5.5 Circuit for modification of charge distribution in series capacitor combination

necting suitable resistors in parallel with each capacitor, but the resulting energy wastage may not always be acceptable.

The charge distribution in a series combination can depend on whether any of the components was previously charged. Consider the arrangement of Fig. 5.5. In the absence of the right-hand battery, equal voltages $v_0/2$ appear across each of the identical capacitors. Suppose however that the key K is closed and then reopened. The equilibrium potential difference across the lower capacitor would now be v_0, and it would carry a charge Cv_0. There would be no potential difference across the upper capacitor and it would be uncharged.

5.2.4   Discharging through resistance

Suppose a charged capacitor C is connected into the circuit of Fig. 5.6. Let the switch be closed at time $t = 0$ to allow the capacitor to begin discharging through the resistor R. At some subsequent time t let the situation be as indicated in the diagram. The charges on the capacitor plates are decaying. The current i in the relation

$$i = \dot{q}$$

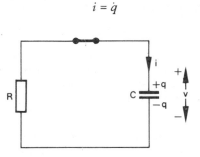

Fig. 5.6 Capacitor discharging through a resistor

will therefore be negative, and the actual direction of flow will be opposite to that indicated in the diagram. The capacitor behaves rather like a source of emf, the magnitude of which falls continuously during the discharge process. The energy stored by the capacitor is meanwhile being converted irreversibly to heat in the resistor.

Kirchhoff's second law gives

$$Ri + \frac{q}{C} = 0.$$

The current i can now be eliminated between the two relations above, giving

$$q = -CR\dot{q}.$$

This is rearranged as

$$dt = -CR \frac{dq}{q},$$

which has then to be integrated. In deriving analytic solutions of physical problems it is often good policy to work with ratios, so that one does not become involved with mathematical functions of dimensional quantities. The constant of integration for the above equation is therefore written as $CR \log_e q_0$, and attached to the right-hand side of the resulting equation. The solution then takes the simple form

$$t = -CR \log_e (q/q_0).$$

Here the symbol e is the base of Naperian logarithms, and q_0 is seen to be the charge on the capacitor at time $t = 0$. The ratio q/q_0 is evidently dimensionless.

Rearrangement of the above equation gives

$$q = q_0 \exp (-t/CR).$$

By setting $i = \dot{q}$ a similar relation is obtained for the current, viz:

$$-i = i_0 \exp (-t/CR),$$

where $i_0 = q_0/CR = v_0/R$. Here i_0 is evidently the magnitude of the initial current, and v_0 is the initial potential difference across the capacitor. Dividing through the relation for q by C gives

$$v = v_0 \exp (-t/CR).$$

Inspection of the relations obtained for q, $-i$ and v shows that all these quantities decay exponentially with time in the manner of Fig. 5.7. Each is reduced to a fraction $1/e$ of its initial value in a time interval

$$\tau = CR.$$

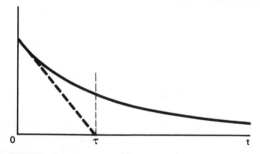

Fig. 5.7 Exponential decay

This quantity is called the time constant of the circuit. If C is in farads and R is in ohms, then τ is in seconds.

It is worth noting that for quantities which decay exponentially with time, the fractional rate of change with time is constant and numerically equal to the reciprocal of the time constant. It is quite easy to show for the above circuit that

$$1/CR = -\dot{q}/q = -\dot{i}/i = -\dot{v}/v.$$

The slope of a tangent to the exponential decay curve illustrated in Fig. 5.7 is therefore

$$-\frac{\text{instantaneous value of the decaying quantity}}{\text{the time constant}}.$$

The tangent to the curve at time $t = 0$ therefore makes an intercept τ on the time axis. It is noticeable that if the initial rate of change of the quantity were maintained, it would reach its final steady value in a time equal to the time constant. This interpretation provides an alternative definition for the time constant which can be extended to circuits containing components which have nonlinear voltage–current relationships and do not therefore exhibit exponential decay characteristics.

5.2.5   Charging through resistance

Consider the arrangement of Fig. 5.8 in which the capacitor C is initially uncharged. Suppose that at time $t = 0$ the key is closed and charging commences. Ultimately the potential difference across the capacitor approaches in magnitude the emf of the source, and current flow will then have virtually ceased.

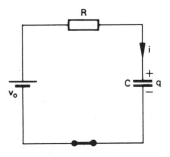

Fig. 5.8 Capacitor charging through a resistor

At some intermediate time t let the situation be as in the diagram. Kirchhoff's second law gives

$$\frac{q}{C} + Ri = v_0.$$

Also

$$i = \dot{q}.$$

Eliminating the current i between these two equations,

$$v_0 - \frac{q}{C} = R\dot{q}.$$

Notice that

$$R\dot{q} = -CR\frac{\mathrm{d}}{\mathrm{d}t}(v_0 - q/C)$$

identically. The differential equation can therefore be rearranged as

$$\frac{\mathrm{d}(v_0 - q/C)}{v_0 - q/C} = -\frac{\mathrm{d}t}{CR}.$$

This easily integrates to the form

$$\log_e \frac{(v_0 - q/C)}{v_0} = -\frac{t}{CR},$$

where the constant of integration has been inserted on the left-hand side as a term $-\log_e v_0$, ensuring satisfaction of the initial condition

$$q = 0 \text{ for } t = 0.$$

Let the final full charge on the capacitor be represented by the symbol q_0. Evidently

$$q_0 = Cv_0,$$

and our equation can now be rearranged as

$$q = q_0[1 - \exp(-t/CR)].$$

Alternatively, dividing through by C, the potential difference across the capacitor at time t is

$$v = v_0[1 - \exp(-t/CR)].$$

The variation with time of both charge and potential difference takes the form shown in Fig. 5.9. Each grows from zero, and levels off at a steady value as t tends to infinity.

Using the relation $i = \dot{q}$, the current i is easily found to take the form

$$i = i_0 \exp(-t/CR),$$

where the current at time $t = 0$ is $i_0 = v_0/R$. This initial value suggests that when the switch is closed the capacitor behaves momentarily as a dead short. This is not an unreasonable analogy, for at that instant, although current is flowing in the circuit, there is no difference in potential between the uncharged plates of the capacitor.

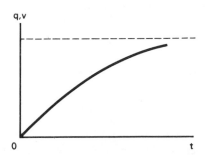

Fig. 5.9 Growth of charge and potential difference during charging of capacitor

The current falls asymptotically to zero at a rate determined by the time constant CR, its behaviour being in every respect the same as in the discharge process encountered previously. The expressions for the charge and potential difference have become rather more complicated as a result of the introduction of the battery into the circuit. But for all the quantities mentioned, the time-dependence during charging is characterised by a factor $\exp(-t/CR)$, which imposes a time constant CR on the variation.

5.2.6 *Multi-loop networks*

Analysis of the transient behaviour of a branched circuit may be time consuming if attempted from first principles, but helpful short-cuts can sometimes be devised. By way of illustration we shall investigate the transient responses of some simple branched circuits.

(*a*) Let us first consider the situation in the circuit of Fig. 5.10 at a time t after closure of the key. The capacitor is taken to be initially uncharged.

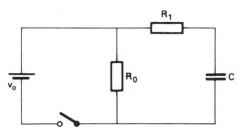

Fig. 5.10

It happens to be particularly easy to predict the behaviour of this circuit. The resistor R_0 conducts a steady current v_0/R_0, and is otherwise without influence, for the source is assumed to have zero internal resistance and is therefore able to maintain a constant difference in potential across the resistor. The branch containing the capacitor is supplied with the same constant difference in potential, so that the charging current is

$$\frac{v_0}{R_1} \left[1 - \exp\left(-t/CR_1\right)\right].$$

The total current being drawn from the battery at any instant is therefore

$$\frac{v_0}{R_0} + \frac{v_0}{R_1} \left[1 - \exp\left(-t/CR_1\right)\right].$$

Suppose the capacitor is allowed to charge fully. If now the key is opened, the character of the circuit is modified, because in effect the battery has been removed. The capacitor discharges through a total resistance $R_0 + R_1$, so that the time constant is $C(R_0 + R_1)$, and the current at any time t measured from the instant of opening the switch is

$$\frac{v_0}{R_0 + R_1} \exp\left[-t/C(R_0 + R_1)\right].$$

(*b*) The charging process becomes rather more complicated if one of the resistors is transferred to the battery loop as in Fig. 5.11.

It is reasonable to suppose that the magnitude of the battery emf is without influence on any time constant in the circuit. If then v_0 were made vanishingly small, the battery could be replaced by a short-circuit, and resistors R_0 and R_2 would form a parallel combination. It appears therefore that the source resistance for charging of C is $1/(1/R_0 + 1/R_2)$, giving a time constant of magnitude $CR_0R_2/(R_0 + R_2)$.

In view of the potential-dividing action of the resistors, the steady difference in potential which is developed ultimately across the capacitor is

$$v_0R_0/(R_0 + R_2) = v_0/(1 + R_2/R_0).$$

The capacitor will be supposed to be initially uncharged. The potential

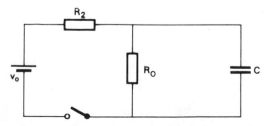

Fig. 5.11

difference across it at any subsequent instant in time must therefore take
the form

$$\frac{v_0}{1 + R_2/R_0}\left[1 - \exp\left(-t\,\frac{R_0 + R_2}{CR_0R_2}\right)\right],$$

for this gives the correct values at times $t = 0$ and $t = \infty$, and embodies also
the right time constant.

Notice that in effect the method of solution involves identifying the
initial and end states of the circuit, and linking them with a transient
response of appropriate time constant.

Expressions for the currents flowing elsewhere in the circuit can
easily be constructed. For example, the current in R_0 is equal to the potential difference developed across the capacitor, divided by R_0.

In analyses of this kind, it is worth checking simple features of the
formulae obtained with a view to testing their validity. It would be found
for example that the current predicted for R_0 would be zero at time $t = 0$.
Is this consistent with the magnitude of potential difference which would
appear across the capacitor at the same instant?

(*c*) The network of Fig. 5.12 combines features of both of the circuits just considered. The capacitor may be supposed to be initially
uncharged, and the battery key is closed at time $t = 0$. The following
method of solution is suggested.

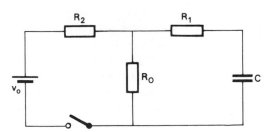

Fig. 5.12

As judged from the terminals of the capacitor, current is being
drawn through the resistor R_1 in series with resistors R_0 and R_2 in parallel.
The effective source resistance is therefore

$$R_1 + 1/(1/R_0 + 1/R_2) = R, \quad \text{say.}$$

When the capacitor is finally fully charged, no current flows in the capacitor branch, and the potential difference across C is then the same as that
across R_0, viz:

$$\frac{v_0R_0}{R_0 + R_2} = v_0/(1 + R_2/R_0).$$

The difference in potential across C grows from zero with time constant CR, so that the value at any time t must be

$$\frac{v_0}{1 + R_2/R_0} \left[1 - \exp\left(-t/CR\right)\right],$$

where R is defined as above.

Using this expression as starting point, it is not difficult to devise relations for any of the various voltages and currents in the circuit.

5.2.7   *Practical capacitors*

The earliest condensers appeared around 1745, and it was in that year that the Leyden jar was invented. Capacitance values of the early condensers were small in relation to volume, but impressive quantities of charge could be stored with the aid of the high voltages generated by friction machines. A great deal of progress has since been made in reducing capacitor dimensions, and miniaturisation continues to be an important preoccupation. Additional desirable features are stability and precision of capacitance values, and reduction of loss (section 6.11).

With a given design geometry, capacitance can be maximised by suitable choice of dielectric. For example, the increase is in the range four to eight times if mica is used instead of air. An overall figure of merit for a dielectric is the permissible energy storage per unit volume. Here an important limitation is dielectric breakdown. In air at normal pressure this occurs when the potential gradient rises to about 3 kV mm^{-1}, but this value can be exceeded in non-gaseous dielectrics. Ohmic conduction is insignificant in most dielectrics, but all exhibit some loss when subjected to the influence of alternating voltages of high frequency.

Most continuously variable capacitors are manufactured in the form of interleaved parallel metal plates. One set of plates is fixed in position, and the other set, which is usually earthed, can be rotated by means of a calibrated head. The plates can be shaped to give a suitable relation between capacitance and angle of rotation. The range of capacitance can be increased by replacing the intervening air with a suitable liquid. Some variable capacitors are cylindrical in construction, and a precision form is available in the form of overlapping coaxial cylinders. Penetration is controlled by a micrometer screw arrangement, which provides an accurately linear control of capacitance over much of the range. Variable capacitance is more often provided in the form of decade boxes, in which fixed capacitors can be selected for parallel connection by rotary switches.

Capacitors designed for use as secondary electrical standards are fixed in value, and have air or good quality solid dielectrics. Ordinary commercial capacitors are basically thin interleaved layers of tinfoil and mica embedded in wax, or silvered mica, or silvered ceramics. The dielectric layers are usually extremely thin. Other insulating materials include waxed paper, glass and plastics.

Electrolytic condensers use aluminium or tantalum foil in contact with a suitable solution. Passage of electric current in the appropriate direction forms a thin insulating oxide film on the metal, the thickness of which may be as little as a few nanometers. Very high stored energies are available in this way, but use is restricted to situations in which a dc polarising voltage can be provided. Electrolytic condensers are commonly employed where an alternating voltage of relatively small amplitude is superimposed on a dc voltage. The maximum permissible voltage is usually prominently marked on the capacitor. It may be as little as a few volts and should on no account be exceeded. The insulating film will only form if the polarising voltage is correctly connected. Such condensers are usually self-healing after breakdown, but the life is shortened if the condenser is subjected to extreme electrical or thermal conditions.

Capacitance can exist between either plate of a condenser and other conductors in the vicinity. These stray effects can be markedly reduced with the aid of an earthed metal screen. This eliminates the unwanted capacitance and gives rise in its place to capacitance-to-earth, which is likely to be stable and generally less troublesome. Some commercial capacitors are prominently marked to show which plate should preferably be attached to earth or to the connection of lower potential.

It is not uncommon for capacitors in electronic equipment to be carrying lethal charges, which may persist long after disconnection from supplies. Electrical shocks have been experienced as much as a whole day after an apparatus has been switched off, and where possible the time constant should be reduced by means of a shunting resistor, or 'bleeder' as it is called.

5.3   Inductance

The phenomenon of electromagnetic induction was first demonstrated in 1831 by Faraday during his most fruitful period of scientific discovery. Two coils wound on a soft-iron ring were connected into separate closed circuits. One of these contained a battery and will be called the primary circuit. The other will be called the secondary circuit. When the current in the primary coil was changed, a current was observed to flow momentarily in the secondary circuit. It is probable the discovery was in part accidental, and that Faraday's intention was to produce interaction with steady current. We now know the effect occurs only with varying current, and recognise Faraday's pair of coils as the first transformer.

5.3.1   *Transformer effect*

Consider the arrangement of Fig. 5.13, in which two coils of wire are connected in separate closed circuits, and placed in close mutual proximity. When the battery key is closed, current will begin to flow in the primary circuit. A ballistic galvanometer (BG) is provided for detection of current in the secondary circuit. In both circuits current-limiting series resistors are provided. It is found that a brief pulse of current flows in the secondary circuit when the key is closed, and again, but in the opposite direction, when it is opened. The directions of these secondary currents are reversed if the observations are repeated with the battery connections interchanged, and their magnitudes are found to be reduced if the two coils are moved apart.

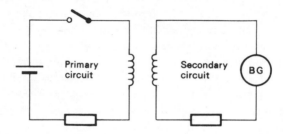

Fig. 5.13 Inductively-coupled circuits

It is important to notice that the secondary current flows only while the primary current is changing. We can suppose that when the current in the primary coil is changing it gives rise to an emf in the secondary coil, and that this in turn drives a current through the closed secondary circuit. The induced emf is distributed through the turns of the secondary, although it is convenient for purposes of analysis to represent it as a localised emf connected in series with the coil.

Experiment indicates that there is a linear connection between the magnitude of this induced secondary emf v_2 and the rate of change with time of the primary current i_1. This suggests the relation

$$v_2 = M_{12} \frac{di_1}{dt},$$

where M_{12} is a constant for a given arrangement of the coils, and is known as the mutual inductance. The form of the equation is consistent with the observation that the sign of v_2 depends on the direction of flow of the primary current, and on whether the primary current is experiencing growth or decay.

If the units of v_2 and $\dfrac{\mathrm{d}i_1}{\mathrm{d}t}$ are volts and amperes per second respectively, then the units of M_{12} are henries. The henry (H) is a rather large unit, and millihenries (mH) and microhenries (μH) are more commonly encountered.

The mutual inductance between two coils is progressively reduced if the separation between them increases. On the other hand it is made very much larger if they are wound on a soft-iron ring, as in Faraday's apparatus.

The phenomenon we have discussed here is the *transformer effect*. It occurs in circuits which are fixed in position, and should not be confused with the related phenomenon known as the *dynamo effect*, which depends for its existence on relative motion.

For coils wound on a ferrous core, the mutual inductance depends on the instantaneous values of the currents in the coils, with the result that circuit analysis is complicated by nonlinearity. Ferrous-cored coils are nevertheless essential if effective coupling is to be obtained at frequencies as low as that of the ac mains.

Transformer-coupling is a rather difficult subject for study, even where relations are strictly linear, and we shall therefore divert our attention temporarily to the simpler but related phenomenon of self induction.

5.3.2 Self induction

The turns of a coil are in general in fairly close proximity to one another, so that a changing current in any one turn induces significant emfs in each of the others. The phenomenon is called *self induction*, and it is obviously an additional manifestation of transformer effect.

An air-cored coil is represented diagrammatically as in Fig. 5.14(*a*); where a ferrous core is provided the modification of Fig. 5.14(*b*) is adopted. It is convenient to treat a self-induced emf as if localised, but its

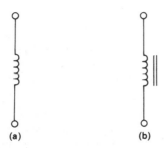

(a)   (b)

Fig. 5.14 (*a*) Air-cored coil, (*b*) iron-cored coil

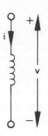

Fig. 5.15 Distributed representation of induced emf

distributed character can be emphasised by the representation used in Fig. 5.15.

It was discovered by Lenz that an induced emf always acts in such a way as to oppose the cause. In the diagram above, the sign of v is to be taken as positive when $\dfrac{\mathrm{d}i}{\mathrm{d}t}$ is positive, in which case the induced emf opposes the growth of current. A self-induced emf is therefore sometimes known as a back-emf, although this name can be misleading. For a decaying current the polarity of the emf is reversed, so that both $\dfrac{\mathrm{d}i}{\mathrm{d}t}$ and v are numerically negative. An abrupt change of current is impossible since it would be opposed by an infinitely-large induced emf.

5.3.3   Coefficient of self inductance

The linear connection noted earlier between the rate of change of current and the magnitude of the induced emf suggests for self induction in a single coil the relation

$$v = L\,\frac{\mathrm{d}i}{\mathrm{d}t}.$$

Here L is a constant for an air-cored coil and is known as the *self inductance*. The sign convention to be used in this equation is illustrated in Fig. 5.15. Rigid adherence to this convention is essential if errors in analysis are not to occur. The dimensions of L are the same as for mutual inductance, so that the unit is again the henry. In the present context one henry is therefore the self inductance of a coil in which an emf of one volt develops when the current is changing at the rate of one ampere per second.

The device is called an inductor, and the property inductance, in uniformity with the terminology for resistors and capacitors.

5.3.4 Stored energy

When the switch in Fig. 5.16 is closed, the current i grows from zero, and flows in the direction indicated by the arrow. Application of Kirchhoff's second law gives

$$v_0 = Ri + L\frac{di}{dt}.$$

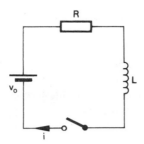

Fig. 5.16 Growth of current in inductor

The sign attached to the last term is positive in conformity with the sign convention already adopted. Notice that for series circuits the terms Ri and $L\frac{di}{dt}$ are additive.

The current is zero immediately after closure of the switch, as the occurrence of any discontinuity in current would be opposed by an infinitely large induced emf.

If the above equation is multiplied through by the current i, a relation is obtained for the instantaneous power delivered by the battery:

$$v_0 i = Ri^2 + Li\frac{di}{dt}.$$

The energy provided in time dt is $v_0 i dt$. The total energy W which is supplied up to time t is obtained by integrating with respect to time from $t = 0$ to $t = t$, giving

$$W = \int_0^t Ri^2 \, dt + \tfrac{1}{2}Li^2.$$

The units of each term in this equation are normally joules, the current being in amperes and the inductance in henries. The integral is recognised as the joule heat dissipated in the ohmic resistance, and the last term must represent energy stored by the inductor.

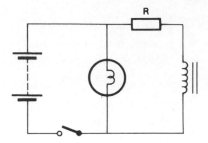

Fig. 5.17 Demonstration of energy storage by an inductor

Storage of energy by a capacitor is a more familiar effect, and the open-circuit nature of the device makes it relatively easy to accept that it is associated with the accumulation of charge. It can be convenient to think of a charged capacitor as possessing *potential energy* in the form of stored charge. In the same spirit the energy of a current-carrying inductor may be regarded as *kinetic* in character, the motion involved being that of the moving charges constituting the current.

A convincing practical demonstration of energy storage by an inductor uses the circuit of Fig. 5.17. The massive iron-cored coil is supplied with a large steady current limited by the rheostat R. The electric lamp is run at a voltage somewhat below normal, so that it draws a relatively small current. The heavy-duty switch is opened and the lamp emits a brilliant flash of light. The experiment demonstrates opposition by the inductor to current change, for the current in the lamp is caused to jump briefly to the same value as was flowing in the coil. The heat and light created momentarily in the lamp filament can originate only from energy stored by the inductor. Of course, a proportion of this energy must be dissipated as joule heat in the rheostat R.

5.3.5 Inductors in series

The instantaneous potential difference developed across the inductors of Fig. 5.18 is

$$v = L_1 \frac{di}{dt} + L_2 \frac{di}{dt} = (L_1 + L_2) \frac{di}{dt}.$$

Evidently the effective inductance of the two inductances is equal to their sum. For any number n of inductors connected in series, the effective inductance is given by the relation

$$L = \sum_1^n L_j.$$

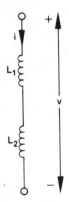

Fig. 5.18 Inductors in series

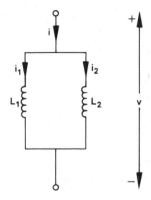

Fig. 5.19 Inductors in parallel

5.3.6 *Inductors in parallel*

The effective inductance L of the parallel combination of inductors of Fig. 5.19 is given by

$$v = L\,\frac{\mathrm{d}i}{\mathrm{d}t}$$

$$= L\,\frac{\mathrm{d}}{\mathrm{d}t}(i_1 + i_2) = L\left(\frac{\mathrm{d}i_1}{\mathrm{d}t} + \frac{\mathrm{d}i_2}{\mathrm{d}t}\right)$$

$$= L\left(\frac{v}{L_1} + \frac{v}{L_2}\right).$$

Evidently

$$\frac{1}{L} = \frac{1}{L_1} + \frac{1}{L_2},$$

and for any number n of inductors connected in parallel the effective inductance is given by

$$\frac{1}{L} = \sum_1^n \frac{1}{L_j}.$$

5.3.7   *Growth in series RL circuit*

Consider the simple circuit of Fig. 5.20, which contains a battery, inductor and resistor. A transient is initiated by closure of the switch.

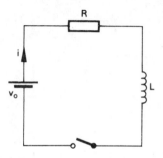

Fig. 5.20 Growth of current in series RL circuit

The current is zero immediately before switching, and will still be momentarily zero immediately after switching. Kirchhoff's second law gives for the situation at any subsequent time t the relation:

$$v_0 = Ri + L\frac{di}{dt},$$

or

$$v_0 - Ri = L\frac{di}{dt}.$$

Notice that

$$L\frac{di}{dt} = -\frac{L}{R}\frac{d}{dt}(v_0 - Ri)$$

identically. The differential equation can therefore be rearranged as

$$\frac{\mathrm{d}(v_0 - Ri)}{v_0 - Ri} = -\frac{R}{L}\,\mathrm{d}t.$$

This can be integrated immediately to the form

$$\log_e \frac{v_0 - Ri}{v_0} = -\frac{Rt}{L},$$

where on the left-hand side a constant of integration $-\log_e v_0$ has been added so that the initial condition $i = 0$ for $t = 0$ may be satisfied. The solution is more usually written

$$i = \frac{v_0}{R}\left[1 - \exp\left(-Rt/L\right)\right],$$

and has the form illustrated in Fig. 5.21. The time constant of the circuit is identified as $\dfrac{L}{R}$, for it is this quantity which determines the rate of decay of the time-dependent term.

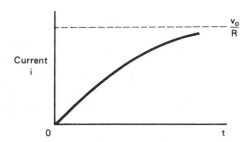

Fig. 5.21 **Time-dependence of growing current in series RL circuit**

As there is initially no current in R, the full battery emf appears across the inductor at the start, and the initial rate of rise of current is given by the relation

$$L\left(\frac{\mathrm{d}i}{\mathrm{d}t}\right)_{t=0} = v_0, \quad \text{or} \quad \left(\frac{\mathrm{d}i}{\mathrm{d}t}\right)_{t=0} = \frac{v_0}{L}.$$

This is independent of the resistance of R.

As the process nears completion, the current approaches the steady value $\dfrac{v_0}{R}$. The final state of the circuit is therefore independent of the value of L.

5.3.8   *Decay in series RL circuit*

When the switch in the circuit of Fig. 5.22 has been closed for some time, a steady current, i_0 say, will be flowing downwards in the inductor. The magnitude of this current is limited by the resistor r. If the inductor L is free of resistance, no current will flow in R, as the inductor will behave as a short-circuit for steady currents. In general L will possess some resistance, so that some flow in R does occur, but this will not affect the character of the analysis which follows.

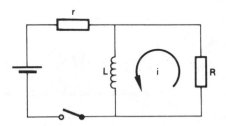

Fig. 5.22 Circuit for initiation of LR decay transient

Let the switch be opened at time $t = 0$. The current in L cannot change discontinuously. The current in R therefore jumps to the value i_0, so that an anticlockwise mesh current of this magnitude will now be flowing in the LR loop.

Kirchhoff's second law gives for the situation at any subsequent instant:

$$0 = Ri + L\frac{\mathrm{d}i}{\mathrm{d}t}.$$

This is first rearranged as

$$\frac{\mathrm{d}i}{i} = -\frac{R}{L}\,\mathrm{d}t.$$

When this is integrated, a constant of integration $-\log_e i_0$ should be added to the left-hand side, so that the initial condition $i = i_0$ can be satisfied. Rearrangement of the resulting equation gives

$$i = i_0 \exp{(-Rt/L)}.$$

The current therefore decays exponentially.

The time constant is L/R. Compare this with the value CR obtained for a resistance-capacitance combination. Evidently R has a quite different influence in the two situations. In the CR circuit R limits the current, and therefore also the rate of decay of the charge on the condenser. In the LR circuit, inspection of the relation

$$L \frac{di}{dt} = -Ri$$

reveals that any increase in R must increase the value of $-\dfrac{di}{dt}$ for a given current i, so that a more rapid decay of current results. In consequence, high resistance gives rise to a large induced emf of relatively brief duration.

If R is made very large, the initial potential difference developed across L is also large. This process is taken to its limit when a circuit containing a current-carrying coil is opened by means of a switch. If the insulation is good, the coil discharges through the stray capacitance between its own turns, but otherwise there may be arcing at the switch contacts and even dielectric breakdown at areas of weakness in the insulation. The circuit shown in the diagram is not susceptible to these problems if R is not made too large, because the LR loop remains intact during switching.

Transient decay is responsible for a minor hazard in circuit continuity testing. If a large inductance such as a secondary winding of a mains transformer is connected between the probes of the ohmmeter, when the user breaks the circuit he may experience momentarily a voltage much greater than that provided by the battery of the instrument.

The transient in a passive CR circuit may be said to be charge-initiated. Any current flowing at the instant of switching is without influence, for it can be changed discontinuously, and subsequent events will be dominated by the charge on the condenser. By contrast, in a passive LR circuit the transient is current-initiated, and charge has no direct effect.

5.3.9 Sign convention in transformer effect

For the coupled circuits of Fig. 5.23, subscripts 1 and 2 denote quantities

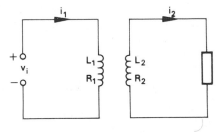

Fig. 5.23 Mutually-coupled circuits

associated with the primary and secondary circuits respectively. A source of constant emf v_i is connected to the primary, and a resistive load is connected to the secondary. The resistance of the primary coil is R_1, and the total series resistance of the closed secondary circuit is R_2. The self inductances of the primary and secondary coils are L_1 and L_2 respectively.

As part of our sign convention for circuits coupled in this way, we shall require the directions of the currents i_1 and i_2 to be represented as in the diagram. Kirchhoff's second law for the primary and secondary respectively will then give the relations

$$v_i = R_1 i_1 + L_1 \frac{di_1}{dt} + M_{21} \frac{di_2}{dt},$$

and

$$0 = R_2 i_2 + L_2 \frac{di_2}{dt} + M_{12} \frac{di_1}{dt}.$$

Here the quantity M_{21} is defined as the emf induced in circuit 1 by unit rate of change of current in circuit 2. M_{12} is defined in a similar way. Notice each of these quantities appears in a term which is *added* to the corresponding $L \frac{di}{dt}$ term. In effect this defines the sign convention to be used for M_{12} and M_{21} in association with the chosen directions for i_1 and i_2.

If the circuit connections to the primary coil are interchanged, the direction of flow of the current in that coil will be reversed for a given direction of flow in the attached circuit. The polarity of the emf induced by that current in the secondary will therefore also be reversed. Furthermore, the direction in which current tends to flow in the external primary circuit because of a given polarity of induced emf in the primary coil will also be reversed. Corresponding reversals occur if alternatively the connections to the secondary are interchanged. The effect of interchanging the connections to either coil is to change the signs of M_{12} and M_{21}. Whereas quantities like resistance, inductance and capacitance are essentially positive for passive circuit elements, a mutual inductance can take either sign, and the sign may be changed by simple interchange of the connections to either coil. Notice that if the connections to both of the coils are simultaneously interchanged, then the signs of M_{12} and M_{21} remain unchanged.

5.3.10 Coefficient of mutual inductance

Let us suppose that the currents in the primary and secondary of the coupled system of Fig. 5.23 are initially zero, and that these are then caused to increase to final steady values i_1 and i_2. The process may be imagined to take place in either of the ways to be described below. For simplicity ohmic resistance is assumed absent. Its presence would not invalidate the argument.

(*a*) Let the primary current i_1 be established first, with the secondary open-circuited. The current could be drawn from a source of variable current. The energy taken up and stored by the primary during this process is $\frac{1}{2}L_1i_1^2$.

Now let the secondary current be increased from zero to a steady value i_2, using a second source of variable current. The emf $M_{21}\dfrac{di_2}{dt}$ induced in the primary meanwhile is supposed to be neutralised by an adjustable source of emf in that circuit, so that the primary current remains steady at the value i_1. The total energy supplied in this second phase is

$$\tfrac{1}{2}L_2i_2^2 + \int M_{21}\frac{di_2}{dt}i_1\,dt,$$

where the latter term is the contribution by the additional source of emf. i_1 is of course constant here. The total energy supplied is

$$\tfrac{1}{2}L_1i_1^2 + \tfrac{1}{2}L_2i_2^2 + \int M_{21}\frac{di_2}{dt}i_1\,dt$$

$$= \tfrac{1}{2}L_1i_1^2 + \tfrac{1}{2}L_2i_2^2 + M_{21}i_2i_1.$$

(*b*) In the following alternative procedure the roles of primary and secondary are reversed. With the primary open-circuited the current i_2 is established, an amount of energy $\frac{1}{2}L_2i_2^2$ being taken up by the secondary coil in the process. Now build up the current i_1 in the primary, with an adjustable emf $M_{12}\dfrac{di_1}{dt}$ provided in the secondary so that the current i_2 can be maintained constant throughout. In this phase the additional energy supplied is

$$\tfrac{1}{2}L_1i_1^2 + \int M_{12}\frac{di_1}{dt}i_2\,dt.$$

The total energy supplied is therefore

$$\tfrac{1}{2}L_2i_2^2 + \tfrac{1}{2}L_1i_1^2 + M_{12}i_1i_2.$$

As the initial and end states are identical, the energies supplied in procedures (*a*) and (*b*) must be equal. Inspection of the above expressions for these energies indicates that

$$M_{12} = M_{21} = M \quad \text{(say)}.$$

This result is a consequence of the Law of Conservation of Energy.

We can now formally define the coefficient of mutual inductance between two coupled circuits simply as the emf induced in one per unit rate of change of current in the other.

5.3.11   *Mutually coupled series coils*

The potential difference between the terminals of the mutually coupled coils of Fig. 5.24 is the sum of the potential differences across each coil. Thus

$$v = \left(L_1 \frac{di}{dt} + M \frac{di}{dt} \right) + \left(L_2 \frac{di}{dt} + M \frac{di}{dt} \right)$$

$$= (L_1 + L_2 + 2M) \frac{di}{dt}.$$

The effective self-inductance of the arrangement is therefore

$$L_1 + L_2 + 2M.$$

The sign of M can be reversed by interchange of the connections to either coil.

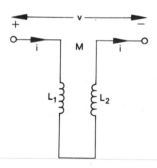

Fig. 5.24 Mutually-coupled series-connected coils

The above arrangement is the basis of the continuously variable self inductor. This is constructed as an air-cored mutual inductor in which the two coils are connected in series, one being movable with respect to the other. The mutual inductance can be adjusted to zero by suitably rotating the axis of the movable coil, and can be made positive or negative.

For larger ranges of self inductance, sets of coils are assembled as a decade box, so that various combinations can be selected by switching.

The layout must be such as to minimise coupling between the various coils.

For low-frequency operation, large inductance values are obtained by the use of laminated ferrous cores. Laminating is essential to minimise energy loss associated with the flow of eddy currents in the core. For higher frequencies the coils have few turns, and the cores are air or a material of low electrical conductivity. In the latter case the mutual inductance can be varied over a rather restricted range by alteration of the core penetration.

For many years the inductor was preferred to the capacitor for the realisation of a primary electrical standard, and for this purpose the mutual inductor was found to be rather easier than the self inductor to design and construct with precision. This point is considered in greater detail in Chapter 12.

5.3.12   *A worked example*

We have seen that the transient behaviour of multi-loop CR networks can be discerned from the characteristics of the basic single-loop CR circuit. The behaviour of a variety of LR networks can be similarly derived from that of the elementary LR series circuit. The principal requirements are recognition of the initial and end states of each given configuration, and of the time constant.

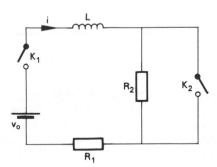

Fig. 5.25

In the following exercise, the time constant and steady-state solution are abruptly changed during the course of a transient response of a single-loop LR series circuit. In Fig. 5.25 both keys are initially open, so that $i = 0$. Key K_1 is closed at time $t = 0$, and key K_2 is closed at time $t = t_1$. We shall calculate the current drawn from the battery at any subsequent instant.

Put

$$R_1 + R_2 = R.$$

We are concerned with growth of current in two phases. During the first phase the series resistance in the circuit is R, and the current grows from zero towards an asymptotic value of v_0/R. At the point P in Fig. 5.26, where $t = t_1$, the current is

$$i_1 = \frac{v_0}{R}\left[1 - \exp\left(-\frac{Rt_1}{L}\right)\right].$$

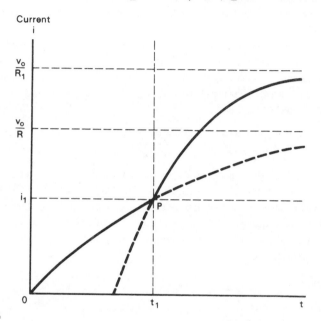

Fig. 5.26

The closure of K_2 reduces the circuit resistance, thereby increasing both the time constant and the ultimate current. Before the form of the second phase can be sketched in the diagram, it is necessary to specify the sign of the discontinuous change in slope which occurs at P. The potential difference developed across the inductor is equal to $v_0 - Ri_1$ just before K_2 is closed, and $v_0 - R_1 i_1$ immediately after. The latter is the larger quantity, so that the slope $\dfrac{di}{dt}$ increases.

The relation between current and time during the second phase is most simply obtained by representing the current as growing from a displaced origin at P. The form of the transient part of the current is

$$i_0\left[1 - \exp\left(-R_1\frac{t - t_1}{L}\right)\right],$$

where i_0 is a constant. Growth develops from an initial current i_1, so i_0 is identified as $\dfrac{v_0}{R_1} - i_1$. A complete expression for the current is therefore

$$i = i_1 + \left(\frac{v_0}{R_1} - i_1\right)\left[1 - \exp\left(-R_1 \frac{t - t_1}{L}\right)\right],$$

or

$$i = \frac{v_0}{R}\left[1 - \exp\left(-\frac{Rt_1}{L}\right)\right]$$
$$+ v_0\left[\frac{1}{R_1} - \frac{1}{R} + \frac{1}{R}\exp\left(-\frac{Rt_1}{L}\right)\right]\left[1 - \exp\left(-R_1\frac{t - t_1}{L}\right)\right].$$

The accuracy of this expression may be checked by confirming that it contains the appropriate time constant, and that it predicts the current correctly for $t = t_1$ and $t \to \infty$.

5.4   Series LCR circuits

In circuits containing both inductance and capacitance two energy storage mechanisms are operating, and the behaviour is in consequence much more varied and complex.

So long as current flows, energy is lost continuously from the system by dissipation in circuit resistance. If the component values are adjusted so that the rate of removal is progressively lessened, a situation develops in which energy can surge to and fro between the inductor and capacitor. The frequency of this oscillatory process depends on the component values, and it is these which determine the principal features of the transient response.

The equation of motion is linear and of the second order, and the solution therefore contains two constants of integration. The values of these constants depend on the initial charge and the initial current, and these determine the finer details of the transient response.

5.4.1   *Equation of charge flow*

Suppose a transient situation exists in the LCR series circuit of Fig. 5.27. The mode of initiation need not concern us at this stage. Kirchhoff's first and second laws give respectively

$$i = \dot{q},$$

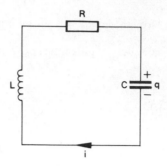

Fig. 5.27 Transient flow of current in series LCR circuit

and

$$0 = L\frac{di}{dt} + Ri + \frac{q}{C}.$$

The current i can be eliminated between these equations, giving the equation of charge flow:

$$0 = L\ddot{q} + R\dot{q} + q/C.$$

Let time-dependence for q of the kind exp $(-\lambda t)$ be assumed, λ being a constant. Then differentiation with respect to time multiplies by $-\lambda$, and the equation takes the equivalent form

$$0 = \left(L\lambda^2 - R\lambda + \frac{1}{C}\right)q.$$

The roots of this quadratic in λ are given by

$$2L\lambda = R \mp \sqrt{(R^2 - 4L/C)}.$$

It is apparent that the form of the solutions for charge and current will depend markedly on the relative magnitudes of R^2 and $4L/C$.

5.4.2   Transient initiation

The presence of a source of emf in the circuit has no influence on the behavioural criterion which we have just encountered. If a battery with emf v_0 is inserted in series with the circuit of Fig. 5.27, with polarity arranged so as to support the flow of current i, then v_0 will appear as a constant additive term on the left-hand side of the equation of charge flow. In section 5.4.8 we shall see that this new term can be removed by the linear substitution

$$q_1 = q - Cv_0,$$

and that its effect is merely to change the value of q at any instant by a constant amount. The battery will in general affect the forms of the initial and end conditions, and in turn the constants of integration. But the time-dependencies of charge and current will be unaffected, and little will be lost in generality if for the moment we ignore the possible influences of batteries.

Relatively simple initial conditions can be devised for which either the charge on the condenser or the current is initially zero. Consider for example the circuit of Fig. 5.28(a). When the key is opened a clockwise current will begin to circulate in the LCR loop, the condenser being initially uncharged. In Fig. 5.28(b) the condenser can be charged by a battery, which is then removed. When the key is closed there will at first be no current flowing, and the transient response which subsequently develops is initiated by the charge on the capacitor.

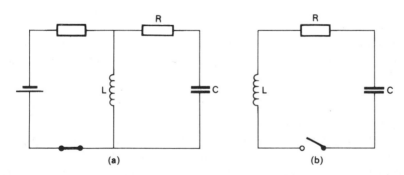

Fig. 5.28 Circuits for generation of (a) **current-initiated transient, and** (b) **charge-initiated transient**

5.4.3 Overdamped circuit

The roots of the equation of charge flow for the series LCR circuit of Fig. 5.27 are real if

$$R^2 > 4L/C.$$

The circuit is then said to be *overdamped*.

Let the roots be λ_1 and λ_2, where

$$2L\lambda_1 = R - \sqrt{(R^2 - 4L/C)}$$

and

$$2L\lambda_2 = R + \sqrt{(R^2 - 4L/C)}.$$

The complete solution is

$$q = A_1 \exp(-\lambda_1 t) + A_2 \exp(-\lambda_2 t).$$

The constants A_1 and A_2 have the dimensions of charge, and their values are determined by the initial conditions in the circuit. The current is

$$i = \dot{q} = -\lambda_1 A_1 \exp(-\lambda_1 t) - \lambda_2 A_2 \exp(-\lambda_2 t).$$

In the event that the charge is initially zero

$$A_1 + A_2 = 0,$$

or if alternatively the current is initially zero

$$\lambda_1 A_1 + \lambda_2 A_2 = 0.$$

In general the motion of the system is characterised by the two time constants $\dfrac{1}{\lambda_1}$ and $\dfrac{1}{\lambda_2}$. In view of the inequality

$$\lambda_1 < \lambda_2,$$

that part of the response which involves the latter time constant decays more rapidly. Incidentally, it is quite easy to show that one or other would be totally absent if either of the conditions

$$i = -q\,\lambda_1, \quad \text{or} \quad i = -q\,\lambda_2$$

happens to be satisfied at time $t = 0$.

If the current is initially zero, the resulting transient is purely charge-initiated. The charge is stationary at time $t = 0$, and decays in the manner of Fig. 5.29(a). The current flows in the opposite direction to that

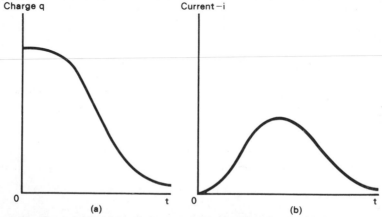

Fig. 5.29 **Charge-initiated transient response in overdamped LCR circuit**

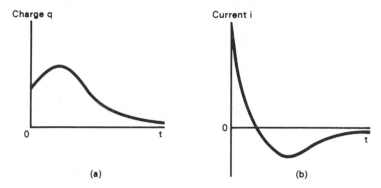

Fig. 5.30 Current reversal in overdamped LCR circuit

indicated in Fig. 5.27. It is zero at $t = 0$ and $t = \infty$, and must therefore pass through a single maximum as indicated in Fig. 5.29(*b*).

The absence of oscillations in the response gives rise to the alternative descriptions *aperiodic* or *deadbeat*. The current will nevertheless experience a change in direction if the sense of flow is initially such as temporarily to reinforce the charge, for in the end state the charge and current must both fall back asymptotically to zero. This situation is illustrated in Figs. 5.30(*a*) and (*b*). The quantity of charge accumulated on the capacitor passes through a maximum as the current changes direction in passing through zero.

5.4.4   Underdamped circuit

If alternatively

$$R^2 < 4L/C,$$

then the roots of the quadratic

$$0 = L\lambda^2 - R\lambda + \frac{1}{C}$$

become complex, and are better represented by the relation

$$2L\lambda = R \mp j \sqrt{\left(\frac{4L}{C} - R^2\right)} = 2L(\alpha \mp j\omega), \quad \text{say},$$

where α and ω are real quantities, and $j = \sqrt{-1}$. Evidently

$$\alpha = R/2L,$$

and

$$\omega = \frac{1}{2L} \sqrt{\left(\frac{4L}{C} - R^2\right)}.$$

A more convenient form for the latter relation is

$$\omega^2 LC = 1 - \alpha^2 LC.$$

The complete solution for the charge is

$$q = A_1 \exp(-\alpha + j\omega)t + A_2 \exp(-\alpha - j\omega)t$$
$$= e^{-\alpha t}[(A_1 + A_2)\cos \omega t + j(A_1 - A_2)\sin \omega t].$$

New constants q_0 and ϕ can be introduced via the substitutions

$$(A_1 + A_2) = q_0 \sin \phi$$

and

$$j(A_1 - A_2) = q_0 \cos \phi.$$

The solution for q then takes the form

$$q = q_0 e^{-\alpha t}\sin(\omega t + \phi).$$

The variation of charge with time is illustrated in Fig. 5.31. The solution is periodic with angular frequency ω. The broken lines indicate the effect on the amplitude of the envelope term $e^{-\alpha t}$.

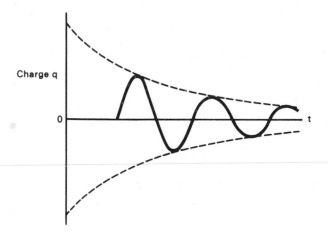

Fig. 5.31 Oscillatory decay of charge

Because of the close relation between charge and current, the current is also oscillatory. When the charge passes through a maximum the energy stored by the capacitor is at a maximum, and when the current passes through a maximum the energy stored by the inductor is at a maximum. The total circuit energy alternates between these two forms with diminishing amplitude.

Non-electrical oscillatory systems can be devised having equations of motion approximating to that of the series LCR circuit. A simple example is a mass attached by a spring to a fixed point. It is because of this analogy that $L\ddot{q}$ is sometimes called the inertial term, $R\dot{q}$ the friction term, and q/C the spring term.

If f is the frequency and T the period of the oscillations,

$$\omega = 2\pi f = \frac{2\pi}{T}.$$

For small damping

$$\alpha^2 LC \ll 1,$$

so that

$$\omega^2 LC = 1, \quad \text{approximately.}$$

In view of the relation

$$i = \dot{q}$$

it follows that

$$i = q_0 e^{-\alpha t} \left[-\alpha \sin(\omega t + \phi) + \omega \cos(\omega t + \phi) \right].$$

We now define new constants i_0 and ϕ_0 via the substitutions

$$\alpha q_0 = i_0 \sin(\phi_0 - \phi),$$

and

$$\omega q_0 = i_0 \cos(\phi_0 - \phi).$$

Then the solution for the current i takes the form

$$i = i_0 e^{-\alpha t} \cos(\omega t + \phi_0).$$

Positive current maxima therefore occur whenever

$$\omega t + \phi_0 = 2n\pi, \quad \text{where} \quad n = 0, 1, 2, \ldots.$$

The amplitude factor takes corresponding values

$$\exp \left[-\alpha \frac{2n\pi - \phi_0}{\omega} \right].$$

The ratio of the amplitudes of successive maxima can be deduced by considering the effect of changing n by 1, and is seen to be

$$\frac{y_n}{y_{n+1}} = \exp \left(\frac{2\pi\alpha}{\omega} \right) = \exp(\alpha T).$$

Therefore

$$\log_e \frac{y_n}{y_{n+1}} = \alpha T = \delta, \quad \text{say,}$$

δ is called the logarithmic decrement of the oscillations. Sometimes the logarithmic decrement is defined in terms of the ratio of successive current maxima of alternating sign, in which case T is replaced by $\dfrac{T}{2}$ in the last relation.

5.4.5 *Critically damped circuit*

If the component values in the series LCR circuit are adjusted so that

$$R^2 = \frac{4L}{C},$$

then

$$\lambda = \frac{R}{2L},$$

and the two roots of the quadratic are indistinguishable.

Suppose that in an overdamped series LCR circuit the magnitude of the quantity $R^2 - 4L/C$ is progressively reduced. λ_2 becomes smaller and λ_1 becomes larger, so that $1/\lambda_2$ becomes larger and $1/\lambda_1$ becomes smaller. Each of these time constants approaches the critical value $2L/R$. The duration of any transient response will be lessened, for in general it is that part of the solution associated with the longer time constant $1/\lambda_1$ which is dominant as the system approaches the steady state.

Of course, if R^2 is made less than $4L/C$ the response will become oscillatory, and quiescence is then approached only after an appreciably greater interval. It is apparent therefore that it is in the condition $R^2 = 4L/C$ that the circuit most rapidly approaches the steady state. It is then said to be *critically damped.* We ought not however to forget that any exponential decay process takes in principle an infinite amount of time to reach true quiescence.

Moving-coil ammeters and voltmeters are generally designed so that the motion of the coil is approximately critically damped (section 4.3.3). Slight underdamping is favoured, so that there is partial overshoot followed by brief damped oscillation about the position of equilibrium deflection. In this way the coil rapidly attains a position approximating to the final equilibrium, the subsequent motion being confined within narrow limits. Most of the damping in such instruments is provided by energy-consuming eddy currents in the conducting former on which the coil is wound. By contrast, insulating formers are used in ballistic galvanometers so that damping is minimised. The coil is set in motion by a brief surge of charge through the windings, and the intention is that as little energy as possible should be lost while the deflection builds up to its first maximum.

With the merging of the roots of the quadratic, it might seem that there will be only one constant in the solution of the equation of motion. But two adjustable constants are essential, as the equation is of the second order. One can make a reasonable guess that the complete solution will contain a factor $\exp(-\lambda t)$, which suggests that a solution of the kind

$$q = f(t) \exp(-\lambda t)$$

should be sought, where the form of $f(t)$ is as yet unspecified. The first derivative of the charge with respect to time is

$$\dot{q} = (f' - \lambda f) \exp(-\lambda t),$$

and the second is

$$\ddot{q} = (f'' - 2\lambda f' + \lambda^2 f) \exp(-\lambda t).$$

These quantities are substituted in the equation of motion

$$0 = L\ddot{q} + R\dot{q} + q/C.$$

The resulting equation can be considerably simplified using the relations

$$R^2 = \frac{4L}{C}, \quad \text{and} \quad \frac{1}{\lambda} = \frac{2L}{R}.$$

After elimination of a common factor $\exp(-\lambda t)$ it reduces easily to

$$f''(t) = 0.$$

When this relation is integrated twice with respect to time, two constants of integration appear, the solution taking the form

$$f(t) = A + Bt.$$

The complete solution for the charge is therefore

$$q = (A + Bt) \exp(-\lambda t).$$

A solution for the current may alternatively be sought, and this can be obtained directly from the last equation by means of the relation

$$i = \dot{q}.$$

The current is found to take the same form as the charge, viz:

$$i = (D + Et) \exp(-\lambda t),$$

where the new constants D and E are related linearly to A and B according to

$$D = B - \lambda A$$

and

$$E = -\lambda B.$$

5.4.6   *Charge decay*

Let the key in Fig. 5.32 be closed at time $t = 0$, the capacitor being initially charged. A charge-initiated transient results, the initial current being zero. Suppose for example that the component values give critical damping. The general solution for the current obtained in the previous section was

$$i = (D + Et) \exp (-\lambda t).$$

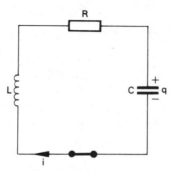

Fig. 5.32 Decay transient in LCR series circuit

This is initially zero if $D = 0$, and then takes the simpler form

$$i = Et \exp (-\lambda t).$$

Also

$$D = B - \lambda A = 0.$$

The expression for the charge was

$$q = (A + Bt) \exp (-\lambda t),$$

where

$$E = -\lambda B.$$

The relations for charge and current can in consequence be written in the forms

$$q = A(1 + \lambda t) \exp (-\lambda t)$$

and

$$i = -\lambda^2 At \exp (-\lambda t).$$

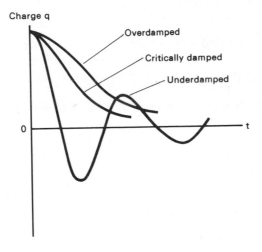

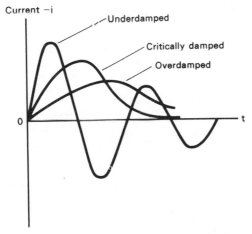

Fig. 5.33 Charge-initiated decay transients in series LCR circuit

The charge decays uniformly to zero without change in sign.

Charge-initiated decay transients are illustrated in Fig. 5.33 for an LCR series circuit in which the component values are adjusted successively to give the three types of damping. The same initial charge is provided for the capacitor in each case. As the initial current is zero, the charge is in each case initially stationary. Notice that the curve of increasing current is always convex upward at the origin. It is worth asking oneself why this should be so. The explanation will be found in the changes which occur shortly after the key is closed in the potential differences across the capacitor and resistor.

5.4.7   *Current decay*

A purely current-initiated transient can be produced in a series LCR circuit with the arrangement of Fig. 5.28(a). When the key is opened the steady current which was flowing in the inductor is immediately established everywhere in the closed LCR loop. Subsequent circuit conditions are effectively those of Fig. 5.32. There is initially no charge on the capacitor.

We shall consider here only the special case of critical damping. In the solutions for charge and current obtained in section 5.4.5 we have

$$A = 0,$$

so that

$$D = B = \frac{-E}{\lambda}.$$

Therefore

$$q = Dt \exp(-\lambda t),$$

and

$$i = D(1 - \lambda t) \exp(-\lambda t).$$

The variations with time of charge and current are illustrated in Fig. 5.34.

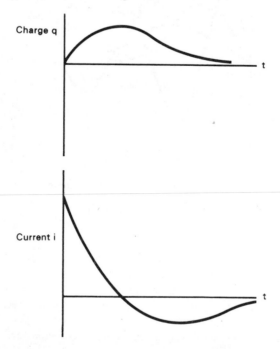

Fig. 5.34 Current-initiated decay transient in critically damped LCR circuit

The constant D is positive, and equal to the initial current. As the capacitor charges up, the current falls to zero at time $t = \dfrac{1}{\lambda}$, reverses direction, and after passing through a maximum in the reverse direction at time $t = \dfrac{2}{\lambda}$ falls asymptotically to zero. The charge on the capacitor reaches a maximum value at time $t = \dfrac{1}{\lambda}$, and then decays.

5.4.8   *Charge growth*

The LCR series circuit of Fig. 5.35 contains a battery of emf v_0. Let the capacitor be initially uncharged. There will be no current flowing immediately after closure of the key, and the resulting transient is initiated entirely by the emf of the battery. After a long period of time current will have ceased to flow, and the potential difference across the capacitor will be equal to v_0.

Kirchhoff's first law gives

$$i = \dot{q}.$$

Applying Kirchhoff's second law we have

$$v_0 = L\,\frac{di}{dt} + Ri + \frac{q}{C}.$$

Elimination of the current gives

$$v_0 = L\ddot{q} + R\dot{q} + \frac{q}{C}.$$

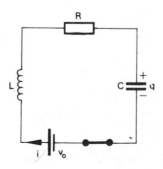

Fig. 5.35 Battery-initiation of transient in series LCR circuit

With the aid of the linear transformation:

$$q_1 = q - Cv_0,$$

we have

$$i = \dot{q} = \dot{q}_1,$$

and

$$0 = L\ddot{q}_1 + R\dot{q}_1 + \frac{q_1}{C}.$$

A simple linear transformation for the charge has eliminated the battery emf from the equation of motion. This suggests that although the battery influences the initial and end conditions in the circuit, the solutions for the charge and current will otherwise resemble those previously obtained in its absence. If for example the values of L, C and R are appropriate for critical damping, then the analysis of section 5.4.5 suggests a solution of the form

$$q_1 = (A + Bt) \exp(-\lambda t).$$

This now becomes

$$q = (A + Bt) \exp(-\lambda t) + Cv_0.$$

The corresponding relation for the current is

$$i = (D + Et) \exp(-\lambda t),$$

where as before

$$D = B - \lambda A,$$

and

$$E = -\lambda B.$$

For the situation under consideration i and q are initially zero, so that

$$D = 0,$$

and

$$A = -Cv_0.$$

Thus

$$q = Cv_0 \left[1 - (1 + \lambda t) \exp(-\lambda t)\right]$$

and

$$i = \lambda^2 Cv_0 t \exp(-\lambda t).$$

In Fig. 5.36 the time-dependencies of charge and current are illustrated for the three types of damping. The initial charge is in each case stationary, as the initial current is zero. In the underdamped circuit the charge overshoots, and then oscillates about the final steady value Cv_0. The current in the critically damped circuit passes through a maximum for

$$t = \frac{1}{\lambda}.$$

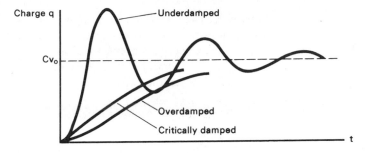

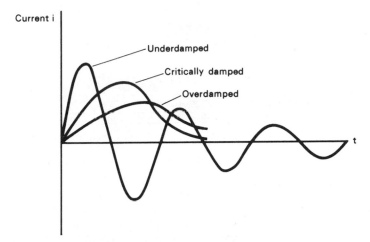

Fig. 5.36 Battery-initiated transients in series LCR circuit

5.5   Calibration of ballistic galvanometer

The ballistic galvanometer is a moving-coil instrument with very low damping and a relatively long period of oscillation, which may amount to as much as 10–20 seconds. If a pulse of charge is caused to flow through the coil in a time which is short compared with the period, the resulting throw is proportional to the magnitude of that charge. The instrument can therefore be employed for the comparison of charges. It may be used for the actual measurement of charge if it is first calibrated by discharging through it a standard capacitor C charged to a known potential difference v. The first throw is measured in terms of the angle of deflection of the coil, or of the movement along a linear scale of a focused beam of light reflected from a mirror attached rigidly to the coil, and should be corrected for damping. The relation between the corrected throw θ and the charge q is

$$\theta = kq = kCv,$$

where k is the charge sensitivity. The correction for damping is affected by the resistance of the circuit in which the galvanometer is connected, and may therefore differ for the conditions of calibration and use.

In an alternative method of calibration, a mutual inductor replaces the capacitor as primary standard. When the switch in Fig. 5.37 is closed, a transient is initiated and a quantity of charge passes through the ballistic galvanometer. Let R_2 be the total series resistance of the secondary circuit. The self inductance L_2 of the secondary circuit is provided largely by the secondary coil, and in part by the coil of the galvanometer. Kirchhoff's second law gives for the secondary circuit the relation

$$0 = R_2 i_2 + L_2 \frac{di_2}{dt} + M \frac{di_1}{dt}.$$

Integrating this with respect to time,

$$0 = R_2 \int i_2 \, dt + L_2 \int di_2 + M \int di_1.$$

If the integration extends through the total duration of the transient response, then the second term on the right-hand side vanishes, because the secondary current is zero at both limits. We are left with the relation

$$R_2 \int i_2 \, dt = -M \int di_1,$$

or

$$q = \frac{M}{R_2} |\Delta i_1|,$$

where q is the total quantity of charge that passes through the galvanometer, and $|\Delta i_1|$ is the net change in the primary current. If a reversing switch is used in the primary circuit to reverse a steady current i, then

$$|\Delta i_1| = 2i.$$

The charge sensitivity is obtained as the ratio of the first throw to the value of q given above.

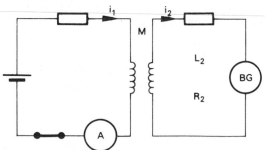

Fig. 5.37 **Calibration circuit for ballistic galvanometer, using standard mutual inductor**

To complete the calibration it is necessary to measure the current i, perhaps in terms of a standard resistance and the emf of a standard cell. The resistance R_2 should be measured in terms of a standard resistance.

The self inductance of the secondary circuit is seen to be without effect. The same is true of resistance and self inductance in the primary circuit. These quantities are not entirely without consequence, however, for all affect the duration of any transient event in the coupled system. It is essential that this be small compared with the period of the galvanometer, so that the instrument can perform its integrating action correctly.

If the effective value of a time-dependent emf in a closed circuit containing a ballistic galvanometer is instantaneously v, then the charge which circulates through the instrument in a given time interval is

$$q = \int \frac{v}{R} \, dt = \frac{1}{R} \int v \, dt,$$

where R is the total resistance of the circuit. The ballistic galvanometer can therefore be regarded as giving a measure of the time-integral of the emf. In the calibration technique which has just been described, the only emf in the circuit which makes any net contribution is of course externally impressed and of magnitude $-M \dfrac{di_1}{dt}$.

5.6   Integration and differentiation

When an electrical signal is applied to any purely ohmic network, its effect is experienced simultaneously at every point within that network, and the responses are wholly free of distortion apart from the inevitable modifications in amplitude. The situation is much less simple in the presence of capacitance and inductance, as the electrical inertia associated with energy storage mechanisms gives rise to delays in response. Thus a finite velocity of propagation exists for electrical disturbances in physically extended non-ohmic networks (section 7.2), and considerable distortion can be caused in circuits of mixed composition. This latter effect may be turned to advantage in various ways. We shall restrict ourselves here to an examination of the integrating and differentiating actions of the simple series CR circuit.

The voltage developed across the resistor R in Fig. 5.38 is proportional to the current i, and the voltage developed across C is proportional to the time integral of that current. Thus v_R is proportional to the time differential of v_C, and v_C is proportional to the time integral of v_R.

Suppose $v_R \gg v_C$, so that $v_R = v_s$, approximately. Let v_s be a signal voltage representing a quantity to be integrated. Then v_C can be used as an output voltage proportional to the time integral of the signal. The condition $v_R \gg v_C$ implies

$$iR \gg \frac{1}{C} \int_0^t i \, dt,$$

or

$$CR \gg \frac{1}{i} \int_0^t i \, dt.$$

The integrating action of the circuit therefore depends on the circuit time constant being long compared with the duration of the event.

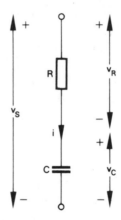

Fig. 5.38 Series CR combination for integration/differentiation

Suppose alternatively that $v_R \ll v_C$. Then $v_C = v_s$, approximately. If the signal voltage v_s represents a quantity to be differentiated, v_R can be used as an output voltage proportional to the time differential of the signal. The condition $v_R \ll v_C$ requires the time constant of the circuit to be short compared with the duration of the event to be differentiated.

In each of the above situations the output voltage is small compared with the signal, and may require a restorative stage of amplification. If C needs to be very large it may be inconvenient or even impossible to provide a capacitor of sufficient size. It is fortunately quite easy to obtain the required integrating action with a suitably modified electronic amplifier.

The effects of differentiation and integration on two specimen waveforms are illustrated in Fig. 5.39. The level output plateaus which result from integration, and which should persist after the event, are likely to sag in practice due to leakage of charge from the capacitor.

What are the principal features of the waveforms produced by integrating or differentiating a continuous sine wave?

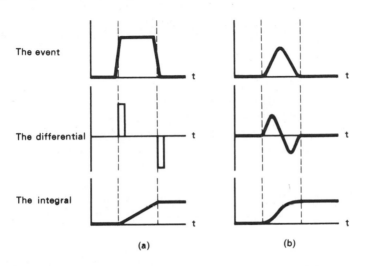

Fig. 5.39 Waveforms produced by differentiation and integration of (a) an approximately rectangular pulse, (b) a rounded pulse

Examples

1. A 20 V battery is able to supply a steady current of 5 A for 100 hours. What capacitor charged to 20 V could store the same amount of energy?

2. A 200 pF capacitor is charged to 100 V and then insulated. The (parallel) plate separation is then increased tenfold. Calculate the potential difference which is now present across the capacitor, and also the electrical energy stored before and after the change.

3. What is the effective capacitance obtained by connecting a 3 μF and two 2 μF condensers as a parallel combination?

4. The potential difference across a charged insulated 4 μF capacitor is 6 V. An uncharged 2 μF capacitor is now connected in parallel with it. Give values for the common potential difference and the total stored energy after sharing.

5. What is the effective capacitance obtained by connecting single 2 μF, 3 μF and 6 μF capacitors as a series combination?

6. A 6 μF and a 3 μF capacitor are connected in series across a 12 V battery. 100 Ω resistors are connected across each capacitor, and after the circuit has reached a steady state both resistors are removed simultaneously. Give values for the charges remaining on each capacitor, and the potential differences across each.

7. Give values for the time constants for
    (a) a 2 μF capacitor discharging through 1 kΩ, and
    (b) an 8 nF capacitor discharging through 5 MΩ.
    If in the latter situation the capacitor had been charged to a potential difference of 500 V, what would be the initial value of the discharge current?

8. A charged capacitor is connected across a resistor. What fraction of the original stored energy will remain after a time equal to the time constant of the circuit?

9. Resistors R_1 and R_2 are connected in series across a capacitor C. A battery of zero internal resistance is connected via an on/off switch in parallel with R_1. What is the time constant
    (a) for the discharging of C when the switch is opened,
    (b) for charging when the switch is closed?

10. A closed circuit is formed by connecting a battery of negligible internal resistance in series with a 2 MΩ resistor, a 1 MΩ resistor and an on/off switch. A 3 μF capacitor is connected in parallel with the 1 MΩ resistor. What is the time constant
    (a) for discharging when the switch is opened, and
    (b) for charging when the switch is closed?

11. A closed circuit is formed by connecting a battery of negligible internal resistance in series with two 2 MΩ resistors and an on/off switch. A 2 μF capacitor is connected in series with a 1 MΩ resistor, and the combination is connected in parallel with one of the two 2 MΩ resistors. What is the time constant
    (a) for discharging when the switch is opened, and
    (b) for charging when the switch is closed?

12. An uncharged 2 μF capacitor is connected in series with a 1 MΩ

resistor, a switch and a 4 V battery. The switch is closed. What is the charge on the capacitor 2 seconds later? If the battery is replaced at this instant by a 2 V battery, what is the charge on the capacitor after a further 2 seconds?

13. What emf is induced in a 50 mH inductor when the current flowing through it is changing at the rate of 4 mA s^{-1}?

14. What is the amount of energy stored by a 50 H inductor carrying a current of 2 A?

15. A 50 H inductor of (series) resistance 100 Ω is connected as a closed circuit with a 1 kΩ resistor. A 110 V dc supply is connected across the resistor. What steady current flows in the resistor? If the supply is suddenly disconnected, what are
    (a) the current in the resistor, and
    (b) the potential difference across the resistor
    immediately after switching? What comment can be made concerning the directions of the currents whose values you have given? What is the amount of energy which is dissipated in the resistor *after switching*?

16. A 25 mH pure inductor is connected across a 5 V source of negligible internal resistance. Give values for
    (a) the initial current, and
    (b) the initial rate of rise of current.

17. An 8 mH inductor having series resistance 1.2 Ω is connected in a closed series circuit with a 12 V source of negligible internal resistance. Give values for
    (a) the initial current,
    (b) the initial rate of rise of current, and
    (c) the final steady current.

18. The effective series resistance of a 36 H self inductor is 60 Ω. A source of emf of negligible internal resistance is connected in series with a 60 Ω resistor and an on/off switch. The combination is connected across the terminals of the inductor. Another 60 Ω resistor is also connected across the terminals of the inductor. Identify the time constants of the circuit with the switch
    (a) closed, and then
    (b) open.

19. A pure self inductor L is connected in a closed series circuit with a resistor R and a source of constant voltage v_0. When a steady current has been established, a second identical resistor R is suddenly connected in parallel with the first. Give an expression for the current drawn from the source at a time $2L/R$ later. Discuss the effect of alternatively connecting the additional resistor
    (a) across the source, and
    (b) across the inductor.

20. What is the magnitude of the emf generated in the open-circuited secondary of a 50 μH mutual inductor if the rate of change of the primary current is 400 A s^{-1}?

21. The self inductances of the primary and secondary of a 100 mH mutual inductor are each 400 mH. What is the total energy stored when the primary and secondary currents are 1 A and 2 A respectively?

22. The self inductance of the primary of an air transformer is 25 mH. The mutual inductance is 20 mH, and the secondary is open-circuited. If a 100 V constant-voltage dc source is suddenly connected across the primary, what maximum emf is generated in the secondary, and at what instant in time relative to the moment of connection?

23. When a 20 mH coil is connected in a closed series circuit with a 10 V source of negligible internal resistance, the resulting transient has a time constant of 4 ms. What is the effective series resistance of the coil? If the mutual inductance between this coil and a second open-circuited coil placed near it is 5 mH, what is the initial magnitude of the emf generated in the latter?

24. A 10 μF capacitor carrying a charge of 100 μC is suddenly connected across the primary of a 5 mH mutual inductor. The self inductance of the primary is 20 mH. If the secondary coil is open-circuited, give values for
    (a) the initial primary current,
    (b) the initial rate of change of primary current, and
    (c) the initial emf developed in the secondary coil.

25. In a series LCR circuit, the capacitance is 1 μF, and the inductor is a 100 μH coil with effective series resistance of magnitude 1 Ω. What

additional resistance should be provided if the circuit is to be critically damped?

26. In a series LCR circuit $L = 0.5$ μH and $C = 400$ pF. What value of R
gives the circuit a natural frequency of oscillation 1% lower than
that obtained with $R = 0$? (An approximate solution is acceptable.)

27. In a series LCR circuit $L = 10$ mH, $C = 1$ μF, and R is adjusted for
critical damping. If the circuit is suddenly completed through a
100 V constant-voltage source, what maximum current flows, and at
what instant after closure of the circuit?

28. In a series LCR circuit, $L = 40$ mH, $C = 1$ μF, and R is adjusted for
critical damping. If the circuit is suddenly completed through a 10 V
constant-voltage source, what is the magnitude of the current 200 μs
later?

29. The sensitivity of a ballistic galvanometer is 100 divisions per μC. It
is connected in the closed secondary circuit of a 50 mH mutual
inductor of total series resistance 100 Ω. What is the throw when a
current 0.5 mA is reversed in the primary coil?

Alternating currents

6.1 Introduction

In the networks studied in earlier chapters, dc sources such as batteries were used, and the resulting current distributions were either steady, or were in the process of settling into the steady state.

When alternating sources are employed, voltage and current vary cyclically with time. The form chosen for the variation depends, of course, on the intended purpose. A great variety of repetitive waveforms finds useful application, and some common examples are illustrated in Fig. 6.1.

Any nonsinusoidal repetitive waveform can be analysed into a set of sinusoidal waveforms of harmonically related frequencies. The process is called *Fourier Analysis*. The component of lowest frequency is the *fundamental*, and components of higher frequency are called *overtones*. If the waveforms of Figs. 6.1(*b*), (*c*) and (*d*) are analysed they are found to contain in addition a constant component, because the displacement which each represents differs on average from zero.

It is usual to employ simple waveforms in both theoretical and experimental investigations of circuit behaviour. Pure sinusoids are favoured, because circuit response often varies markedly with frequency. Such waveforms are fortunately readily available from the ac mains supply and electronic generators.

In Fig. 6.2 a sinusoidally varying quantity

$$v = v_0 \cos (\omega t + \phi)$$

is shown plotted as a function of the time t. The factor $\cos (\omega t + \phi)$ alternates regularly between maximum and minimum values $+1$ and -1 respectively, so that v takes corresponding values $+v_0$ and $-v_0$. The quantity v_0 is called the *amplitude* or *peak* value, and is indicated in the diagram.

If ωt increases by 2π radians from an initial value of ωt_1, then the factor $\cos (\omega t_1 + \phi)$ becomes

$$\cos (\omega t_1 + 2\pi + \phi) = \cos (\omega t_1 + \phi),$$

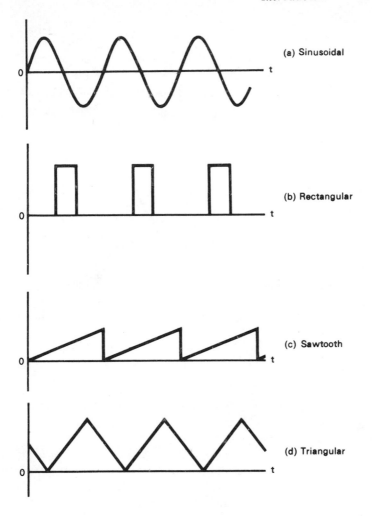

Fig. 6.1 Examples of repetitive waveforms

so that v returns to its original value. It follows that v passes through a complete cycle of variation in a time T given by

$$\omega\,(t_1 + T) = \omega t_1 + 2\pi,$$

so that

$$\omega T = 2\pi,$$

or

$$T = 2\pi/\omega.$$

T is called the *period* of the alternation.

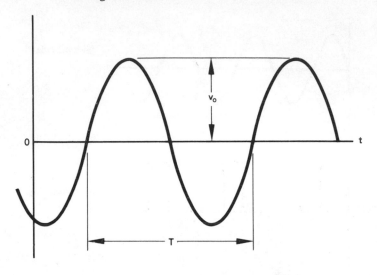

Fig. 6.2 Sinusoidal waveform

The number of complete cycles in unit time is the frequency f, so that

$$f = 1/T.$$

It follows that

$$\omega = 2\pi f.$$

ω is known as the *pulsatance*, or *angular frequency*. As the units of ωT are radians, the units of ω are radians per second. The units of f are cycles per second, or hertz, usually abbreviated to Hz.

The quantity $\omega t + \phi$ is the *phase angle* of the waveform for a given value of time t. ϕ is called simply the *phase constant*. Any alteration in ϕ moves the entire waveform bodily along the time axis.

The quantity $2v_0$ is called appropriately the *peak-to-peak variation* of the waveform.

6.2 Resistive networks

6.2.1 *Current—voltage relation*

Consider the situation of Fig. 6.3, in which a resistor R is connected to a source of alternating emf, of the form

$$v = v_0 \cos \omega t.$$

For an ohmic resistor the relation between current and voltage does not

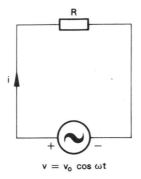

$$v = v_0 \cos \omega t$$

Fig. 6.3 Resistor connected to ac source

change with the magnitude of either, and the instantaneous current is therefore

$$i = \frac{v}{R} = \frac{v_0}{R} \cos \omega t = i_0 \cos \omega t,$$

where the amplitude of the current is

$$i_0 = \frac{v_0}{R}.$$

The variations with time of voltage and current are seen to be determined in each case by the same factor $\cos \omega t$. Their waveforms therefore differ only in amplitude. The resulting identity of phase constant means, for example, that each is zero at the same instant, and that each experiences a maximum at the same instant. The voltage and current are said to be 'in phase'.

6.2.2 Power and energy

For a pure resistor, with sinusoidal waveforms the mean values of potential difference and current are obviously zero, in view of the symmetry of these quantities about the time axis. However, the instantaneous rate of supply of energy is

$$i^2 R = \frac{v^2}{R} = vi,$$

and is always positive. It depends, for a given resistor, on the mean-square current, rather than on the mean current. Now

$$i^2 = i_0^2 \cos^2 \omega t,$$

and the value of this quantity averaged over one cycle is

$$\overline{i^2} = \frac{1}{T} \int_0^T i_0^2 \cos^2 \omega t \, dt.$$

This is easily shown to be equal to $\frac{i_0^2}{2}$, by first converting the integrand to a function of the double angle. It is worth remembering that the average value of $\sin^2 \omega t$ and $\cos^2 \omega t$ over an integral number of half periods is in each case $1/2$. The same is true for $\sin^2 (\omega t + \phi)$ and $\cos^2 (\omega t + \phi)$.

The mean rate of supply of energy P will be called the electric power. In view of the above considerations this is given by

$$P = i_0^2 \frac{R}{2} = \frac{v_0^2}{2R} = \frac{v_0 i_0}{2}.$$

The total energy supplied in time t is

$$\frac{i_0^2 R}{2} t = \frac{v_0^2}{2R} t = \frac{v_0 i_0}{2} t.$$

Here the interval t must cover a whole number of cycles, or the expressions for the mean rate of supply of energy are not applicable.

6.2.3   Root-mean-square quantities

For a given circuit resistance R, the electric power has been seen to depend on the mean-square current $\overline{i^2}$ and the mean-square voltage $\overline{v^2}$. This conclusion is valid for any waveform. In the context of electric power, therefore, the effective current and voltage are the root-mean-square or rms values.

For sinusoidal waveforms it was found

$$\overline{i^2} = \frac{i_0^2}{2}.$$

Similarly

$$\overline{v^2} = \frac{v_0^2}{2}.$$

Thus the rms values are

$$i_{rms} = \sqrt{(\overline{i^2})} = \frac{i_0}{\sqrt{2}},$$

and

$$v_{rms} = \sqrt{(\overline{v^2})} = \frac{v_0}{\sqrt{2}}$$

Inspection of the dc relations for the power

$$P = i^2R = \frac{v^2}{R} = vi$$

shows that these are valid for repetitive waveforms if rms values are used. It is standard practice to employ rms values in alternating-current work, and a given magnitude for a voltage or current can safely be assumed to be rms unless there are clear indications to the contrary. It would be normal, for example, to calibrate the output controls of a signal generator in terms of rms voltages. But notice that the intensity of the shock obtained from the 240 V ac mains supply is associated with the peak value of $240\sqrt{2}$ V = 340 V approximately, rather than the rms value.

The ratio of maximum to rms value is called the *crest, peak* or *amplitude factor*. For nonsinusoidal waveforms the value differs from $\sqrt{2}$. Another waveform parameter is the *form factor*. This is the ratio of the rms to the *half-cycle-average* value, and is $\pi/2\sqrt{2}$ for a sine wave.

6.2.4   Series and branched circuits

Kirchhoff's first and second laws for electric currents depend in that order on the laws of conservation of charge and energy. They are therefore no less valid for alternating currents, although their application is then sometimes less simple and direct.

An important deduction from the first law is that in an unbranched series circuit the current is everywhere the same at a given instant in time. This will be true for 'mixed' circuits containing inductance and capacitance as well as resistance. It will not be true in certain situations which need not concern us greatly at this stage, for example in the insulating dielectric of a condenser, or in circuits where the linear dimensions are so great that the finite time of transit of electrical disturbances is not negligible compared with the period of the ac current.

Imagine now a resistive network, in which a steady dc current distribution exists. Suppose for simplicity there is only one battery in the arrangement. The magnitude of each branch current will be proportional to the emf of the battery. If the emf is doubled, each current will double, and so on. If the original emf can be caused to vary with time, all the currents will vary with time in the same way as the emf, and all will keep in step with the emf. This means that the laws for steady currents in resistive networks can be used without modification for time-dependent currents. Currents in a pair of resistances connected in parallel will divide in inverse proportion to each resistance, the familiar summation law for effective resistance will apply for resistances connected in series, and the reciprocal law for resistances connected in parallel.

Thus for series ohmic combinations

$$R = \sum_1^n R_j,$$

and for parallel ohmic combinations

$$G = \sum_1^n G_j,$$

where G_j is the conductance of the jth shunting branch, and G is the total conductance.

In dc circuits resistance was defined as the ratio of potential difference to current. For ohmic ac circuits it could therefore be thought of as the ratio of the instantaneous potential difference to the instantaneous current. As rms values are favoured for energy and power relations, it is customary to employ instead the relation

$$R = \frac{v_{rms}}{i_{rms}}.$$

This is sometimes replaced by the equivalent quantity $\frac{v_0}{i_0}$, the ratio of peak values.

6.3   AC instruments

It would be difficult, except in very low frequency work, to read the instantaneous value of an alternating current or voltage with an instrument utilising mechanical movement. Most direct-reading ac ammeters and voltmeters provide nonlinear time-averaging, and the response depends in consequence on the waveform, and to a lesser extent on the fundamental frequency.

In the hot-wire ammeter, the current to be measured is passed through a wire under tension. The joule-heating effect causes it to stretch, and the increase in length is observed in terms of the rotation of a pointer. The rate of supply of heat energy is i^2R, so that the deflection is a function of the rms value of the current, regardless of waveform. The resistance of the instrument may vary somewhat with the temperature rise experienced. Like the other instruments described here, the hot-wire ammeter is a *trarsfer instrument* in that it can be used equally well for dc and ac current. The sensitivity of an ammeter can be defined as the ratio of the magnitude of the response to the current producing it. The sensitivity of the hot-wire ammeter is rather low.

The thermocouple ammeter is a modification of the hot-wire instrument. One junction of a thermocouple is mounted close to the heated wire, the resulting thermoelectric emf being detected by a conventional moving-coil instrument. The response is approximately proportional to the square of the rms current. The arrangement has the unique advantage of operating satisfactorily at frequencies as high as 100 megahertz.

Dynamometer-type instruments differ from the conventional moving-coil type in that a fixed second coil replaces the fixed permanent magnet. The fixed and moving coils are connected in series, and the resulting mutual forces are proportional to the square of the instantaneous current. The deflection of the moving coil under the influence of a controlling spring gives therefore a measure of the square of the rms current. The instrument is accurate at low frequencies, but coil inductance and other factors inhibit its use at frequencies above a few hundred hertz. It is used more commonly in modified form as a wattmeter.

Larger mechanical forces are produced in the presence of soft iron. In one form of soft-iron instrument a force of attraction develops between two pieces of soft iron situated inside a coil carrying the current to be measured. One piece of iron is fixed and the other movable; the deflection of the latter is controlled by a spring and observed with a pointer-and-scale arrangement. As with dynamometers, the response depends on the rms value of the current, and use is restricted to frequencies below a few hundred hertz.

For the greatest sensitivity, a moving coil instrument of the d'Arsonval type is used. In Fig. 6.4 the current is rectified by a single diode in series with the coil of the instrument. The response depends very much on the form factor. The diode and instrument must be shunted by a low resistance, so that the current to be measured is neither impeded nor distorted. A better arrangement uses the rectifier bridge of Fig. 6.5, in which both half-cycles are effective, and the overall behaviour is approximately ohmic. This type of instrument is very common. In an alternative form a high resistance is introduced in series with the galvanometer, and a capacitor is connected across the combination. The small current flowing through the galvanometer is proportional to the approximately steady

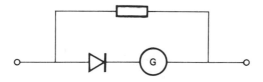

Fig. 6.4 Moving-coil instrument with series diode

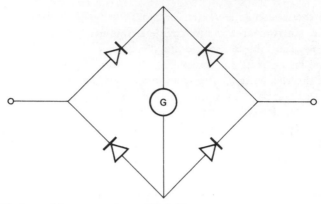

Fig. 6.5 Moving-coil instrument in rectifier bridge

voltage developed across the capacitor, and this in turn is nearly equal to the peak value of the voltage developed across the instrument.

In general any current-measuring instrument can be converted into a voltmeter by connecting a high resistance in series. The response will still depend in the same way on the current, which in turn is approximately proportional to the applied voltage. Thus an instrument which gives a response proportional to the rms current can be used to measure rms voltage.

Differing instruments which have been calibrated under identical electrical conditions give identical readings when subjected to these same calibration conditions. Discrepancies appear if the waveform or frequency are changed. This is because not all the instruments measure the same thing.

6.4   Capacitance

In the circuit of Fig. 6.6, the capacitor C is connected to an ac source of

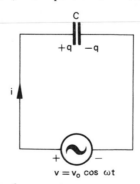

Fig. 6.6 Capacitor connected to ac source

emf. Instantaneous values of charge, current and voltage are indicated on the diagram. An expression can be derived for the current with the aid of the relations (section 5.2.1):

$$i = \frac{dq}{dt} = C\,\frac{dv}{dt}.$$

Here

$$v = v_0 \cos \omega t.$$

Therefore

$$i = -\omega\,Cv_0 \sin \omega t,$$

or

$$i = i_0 \cos(\omega t + \pi/2),$$

where

$$i_0 = \omega C v_0.$$

i_0 is the peak current, and is seen to be proportional to the capacitance and to the angular frequency. The phase of the current is advanced in time by $\pi/2$ radians relative to the applied voltage. The current and voltage are said to be *in quadrature*. Their time-dependencies are illustrated in Fig. 6.7. The relation

$$q = Cv$$

ensures incidentally that the stored charge is always in phase with the voltage.

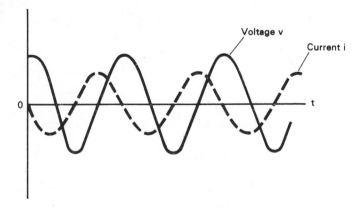

Fig. 6.7 Voltage and current waveforms for a capacitor

The difference in phase between the voltage and current means that their ratio is itself time-dependent, and it therefore lacks precise significance. However, the relations which we have obtained for the capacitor give

$$\frac{v_0}{i_0} = \frac{v_{rms}}{i_{rms}} = \frac{1}{\omega C},$$

and this quantity is independent of time. The ratio of rms voltage to rms current is called the *impedance*. This is a more general term than resistance, and includes it as a special case. The units are of course ohms. Notice that the impedance diminishes as the capacitance and angular frequency increase.

The instantaneous rate of supply of energy to the capacitor is

$$vi = v_0 i_0 \cos \omega t \cos (\omega t + \pi/2)$$

$$= -\frac{v_0 i_0}{2} \sin 2\omega t$$

$$= v_{rms} i_{rms} \cos (2\omega t + \pi/2).$$

This is zero on average, a result which should not be unexpected, because the capacitor lacks any mechanism for converting electrical energy to non-electrical forms. It is, of course, capable of storing energy. In fact, energy surges back and forth cyclically between the capacitor and the supply, the amount stored reaching a maximum twice for each cycle of the applied voltage. Why twice?

At mains frequency the impedance of a 10 μF capacitor is about 300 ohms, so that for the majority of common capacitor sizes the impedance would be higher than this. At a frequency of 50 MHz, however, the same impedance would be obtained with a 10 pF capacitor.

6.5   Inductance

For the circuit illustrated in Fig. 6.8, Kirchhoff's second law gives

$$v = v_0 \cos \omega t = L \frac{di}{dt}.$$

On rearrangement, and integration with respect to time, this becomes

$$i = \frac{v_0}{\omega L} \sin \omega t,$$

or

$$i = i_0 \cos (\omega t - \pi/2),$$

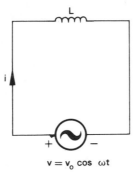

Fig. 6.8 Inductor connected to ac source

where

$$\frac{v_0}{i_0} = \frac{v_{rms}}{i_{rms}} = \omega L .$$

This last quantity is the impedance of the inductor, and it increases both with angular frequency and with inductance. Comparison of the relations given above for voltage and current shows that these quantities are in quadrature, the current lagging the voltage by $\pi/2$ radians. This situation is illustrated in Fig. 6.9.

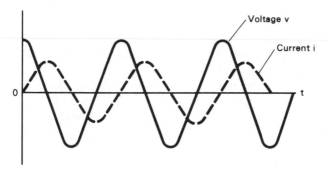

Fig. 6.9 Voltage and current waveforms for an inductor

The instantaneous rate of supply of energy to the inductor is

$$vi = v_0 i_0 \cos \omega t \cos (\omega t - \pi/2)$$

$$= \frac{v_0 i_0}{2} \sin 2\omega t$$

$$= v_{rms} i_{rms} \cos (2\omega t - \pi/2).$$

This quantity is zero on average, and the same remarks can be made here as appeared in this context for a capacitor. Like the capacitor, a pure inductor is capable of storing energy but not of dissipating it.

At mains frequency the impedance of a 1 H inductor is about 300 ohms. At 50 MHz the same impedance would be obtained with only 1 μH.

6.6   Linearity

For the three types of circuit component so far considered, the voltage–current relation is one of strict proportionality. If, for example, the rms voltage is doubled, then so also is the rms current. Behaviour of this kind will be called *linear*.

Any circuit made up as a combination of such components is also linear. Furthermore, by suitable choice of circuit and component sizes it is possible to produce any desired difference in phase between voltage and current.

6.7   Power factor

Suppose an ac source is connected to a linear circuit. Let the supply voltage be $v_0 \cos \omega t$, and the current $i_0 \cos (\omega t - \phi)$. Then the instantaneous rate of supply of energy is

$$v_0 i_0 \cos \omega t \cos (\omega t - \phi) = \frac{v_0 i_0}{2} [\cos (2\omega t - \phi) + \cos \phi].$$

It is noticeable that the first term on the right-hand side alternates at a frequency equal to double that of the source, so that the average value over an integral number of half-cycles of the supply is zero. In consequence, the net rate of supply of energy depends only on the final term. The power is therefore

$$P = \frac{v_0 i_0}{2} \cos \phi = v_{rms} i_{rms} \cos \phi.$$

This is the true wattage, unlike the quantity $v_{rms} i_{rms}$ which is known as the apparent wattage. The factor $\cos \phi$ is called the *power factor*.

The apparent wattage is in general greater than the true wattage, so that for ac circuits the power cannot be estimated merely by measuring the voltage and current supplied.

The power factor is of course unity for resistors, for which $\phi = 0$. It is zero for pure capacitors and inductors, and the current supplied to these is sometimes said to be 'wattless'.

6.8   Basic ac circuit analysis

When an ac voltage is applied to a linear circuit containing inductance, capacitance and resistance, the waveforms of the voltages and currents in the various branches will be everywhere similar to the waveform of the supply, although the mixed character of the components gives rise to phase differences.

For example, in the LCR series combination of Fig. 6.10, the current i is common to all three components. The potential difference across R is in phase with the current, and the potential differences across L and C respectively lead and lag the current by $\pi/2$ radians. The applied potential difference is equal to the sum of all these.

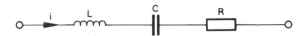

Fig. 6.10 Series LCR combination

Even for a simple circuit like this, any theoretical investigation of behaviour in terms of sinusoidal dependencies on time would be very time consuming. The alternative technique described below allows the exclusion of factors involving time-dependence, so that attention can be concentrated on amplitude and phase relations.

6.8.1   The complex exponent

Let the alternating voltage $v_0 \cos \omega t$ provided by a source in a network be replaced for purposes of analysis by the quantity $v_0 \exp j\omega t$, where $j = \sqrt{-1}$. In view of the identity

$$v_0 \exp j\omega t = v_0 \cos \omega t + j v_0 \sin \omega t,$$

a fictitious imaginary voltage $j v_0 \sin \omega t$ has been added to the original real voltage $v_0 \cos \omega t$. The effect is as if a second source of emf were connected in series with the actual source. In linear circuits the current or voltage distributions associated with two sources exist quite independently of each other, and do not interact in any way. If therefore the response of a linear circuit is investigated for an applied voltage $v_0 \exp j\omega t$, the real distribution can always be extracted by simply ignoring all imaginary terms.

The introduction of the complex exponent greatly simplifies ac circuit analysis. It ensures that time-dependencies of voltages and currents are contained in a common multiplying factor $\exp j\omega t$. This can then be eliminated by cancellation, regardless of any phase differences between the quantities concerned. Suppose, for example, that potential differences $v_1 \cos \omega t$ and $v_2 \cos(\omega t - \pi/4)$ develop across two components in a circuit. Using the complex exponent, these take the forms $v_1 \exp j\omega t$ and $v_2 \exp j(\omega t - \pi/4)$, and a common factor $\exp j\omega t$ can be extracted.

When the complex exponent is employed, circuit relations may contain a lot of complex algebra, and this can seem, at first, a heavy price to pay. The novice finds it difficult to reconcile complex currents and voltages with the readings observed on ammeters and voltmeters. But these readings are in each case a measure of only one feature of an alternating quantity. The relationship which exists between a complex current or voltage and the quantity which it represents is in fact more precise.

6.8.2   Resistance

Consider a simple circuit in which a source voltage $v_0 \cos \omega t$ is applied to a pure resistor R. Using the complex exponent, the source voltage is $v_0 \exp j\omega t$, and conditions are as in Fig. 6.11. Solving in the usual way to find the current i, we have

$$i = \frac{v}{R} = \frac{v_0}{R} \exp j\omega t = i_0 \exp j\omega t \quad \text{(say)}.$$

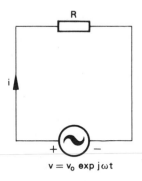

$$v = v_0 \exp j\omega t$$

Fig. 6.11 Resistor with complex applied voltage

Rearrangement of these relations gives

$$\frac{v}{i} = \frac{v_0}{i_0} = \frac{v_{\text{rms}}}{i_{\text{rms}}} = R.$$

The instantaneous current i is seen to be complex. v will be called the *complex voltage,* and i the *complex current.* The actual current can be extracted as the real part of $i_0 \exp j\omega t$, that is, $i_0 \cos \omega t$, since $i_0 \left(= \dfrac{v_0}{R} \right)$ is real.

6.8.3 *Capacitance*

In the circuit of Fig. 6.12, the current in the capacitor is given by

$$i = \dot{q} = C\dot{v} = C \frac{\mathrm{d}}{\mathrm{d}t}(v_0 \exp j\omega t)$$

$$= j\omega C v_0 \exp j\omega t = \omega C v_0 \exp j(\omega t + \pi/2),$$

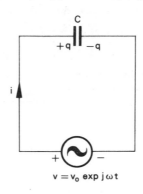

$$v = v_0 \exp j\omega t$$

Fig. 6.12 Capacitor with complex applied voltage

since $j = \exp j\pi/2$. The real part of this last expression for the current is $\omega C v_0 \cos(\omega t + \pi/2)$. It is seen to have the correct amplitude and phase in relation to the applied voltage. From the above relations,

$$i = j\omega C v_0 \exp j\omega t = j\omega C v = i_0 \exp j\omega t, \quad \text{say.}$$

Evidently

$$\frac{v_0}{i_0} = \frac{1}{j\omega C} = \frac{v}{i}.$$

The significance of these last relations will be discussed in the next section.

6.8.4 *Complex impedance*

The quantity $\dfrac{v}{i}$ relates the instantaneous voltage and current, and is called the *complex impedance*. It is generally represented by the symbol Z. The factor $\exp j\omega t$ is common to both v and i, and is therefore absent from their ratio. The complex impedance is real for a pure resistor, and imaginary for a pure capacitor.

Let the voltage developed across a circuit be $v_0 \cos \omega t$, and let the current drawn be $i_0 \cos (\omega t - \theta)$. Then the impedance is

$$\frac{v_{rms}}{i_{rms}} = \frac{v_0}{i_0},$$

and the current lags the voltage by angle θ. Using the complex exponent technique, the source voltage would be represented instead as $v = v_0 \exp j\omega t$. Suppose the solution for the complex current then takes the form $i = a_0 \exp j\omega t$. The real part of this is to be identical with the actual current $i_0 \cos (\omega t - \theta)$. This is the real part of $i_0 \exp j (\omega t - \theta)$, so

$$a_0 = i_0 \exp (-j\theta).$$

a_0 is called the complex amplitude of the current. In conformity, v_0 is generally called the complex amplitude of the voltage, although in the above it has been treated as a real quantity.

The complex impedance is

$$Z = \frac{v}{i} = \frac{v_0 \exp j\omega t}{a_0 \exp j\omega t} = \frac{v_0}{a_0}.$$

Thus

$$|Z| = \left| \frac{v_0}{a_0} \right| = \frac{v_0}{|a_0|} = \frac{v_0}{i_0},$$

since

$$|a_0| = |i_0| \, |\exp (-j\theta)| = i_0.$$

Therefore

$$|Z| = \frac{v_{rms}}{i_{rms}},$$

which is recognised as the impedance. Also

$$\arg Z = \arg \frac{v_0}{a_0} = \arg v_0 - \arg a_0 = \arg v_0 + \theta.$$

As the argument of the real quantity v_0 is zero, therefore

$$\arg Z = \theta.$$

Remember that θ is the *lag* of current behind voltage.

These last relations are of great importance. They enable the ratio of the rms voltage to the rms current and the phase relation between voltage and current to be worked out from a knowledge of the complex impedance.

For example, the complex impedance of a capacitor was found to be $1/j\omega C$. It follows that

$$\frac{v_{\text{rms}}}{i_{\text{rms}}} = \left| \frac{1}{j\omega C} \right| = \frac{1}{\omega C},$$

and

$$\theta = \arg \frac{1}{j\omega C} = \arg(-j) = -\pi/2,$$

confirming that the current leads the voltage by $\pi/2$.

6.8.5   Inductance

For the circuit of Fig. 6.13, Kirchhoff's second law gives

$$v = v_0 \exp j\omega t = L \frac{\mathrm{d}i}{\mathrm{d}t}.$$

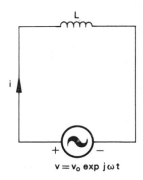

L

i

$v = v_0 \exp j\omega t$

Fig. 6.13 Inductor with complex applied voltage

This is easily rearranged and integrated, giving

$$i = \frac{v_0}{j\omega L} \exp j\omega t = \frac{v}{j\omega L}.$$

The complex impedance of the inductor is

$$\frac{v}{i} = Z = j\omega L.$$

It follows that

$$|Z| = \frac{v_{\mathrm{rms}}}{i_{\mathrm{rms}}} = \omega L,$$

and

$$\theta = \arg Z = \frac{\pi}{2},$$

in agreement with the earlier analysis involving sinusoidal dependence on time (section 6.5).

6.9   Series LCR circuit

For the circuit of Fig. 6.14, Kirchhoff's second law gives

$$v = Ri + L\frac{\mathrm{d}i}{\mathrm{d}t} + \frac{q}{C},$$

where $q = \int i\,\mathrm{d}t$. Here v and i are the instantaneous applied voltage and current respectively. With the voltage represented as $v_0 \exp j\omega t$, v and i will each be of complex form.

The current alternates at the same frequency as the applied voltage, and will therefore contain a multiplying factor $\exp j\omega t$. It follows that

$$\frac{\mathrm{d}i}{\mathrm{d}t} = j\omega i, \quad \text{and} \quad \int i\,\mathrm{d}t = \frac{i}{j\omega}.$$

The initial equation becomes

$$v = Ri + j\omega Li + \frac{i}{j\omega C}.$$

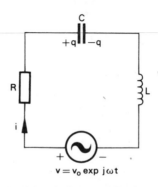

Fig. 6.14 Series LCR circuit with complex applied voltage

The terms on the right-hand side are the complex voltages developed across each of the three circuit components.

It follows immediately that

$$\frac{v}{i} = R + j\omega L + \frac{1}{j\omega C}.$$

The right-hand side is now the sum of the individual complex impedances. The left-hand side is the complex impedance presented by the circuit to the generator. Evidently, for a series circuit the following relation holds

$$Z = \sum_{1}^{n} Z_j,$$

where Z is the effective complex impedance of the complex impedances $Z_1, Z_2, \ldots, Z_j, \ldots, Z_n$ connected in series. This is a generalisation of the addition rule for resistances in series.

6.9.1 Resistance and reactance

Let R and X be the real and imaginary parts of a complex impedance Z. Then

$$Z = R + jX.$$

R is called the *resistance* of Z, as might be expected. X is called the *reactance*.

For the series LCR circuit of Fig. 6.14,

$$R = R, \quad \text{and} \quad X = \omega L - \frac{1}{\omega C}.$$

6.9.2 Argand diagram

It is often helpful to represent complex impedances as points in an Argand diagram. In Fig. 6.15, Z is plotted with Cartesian coordinates (R, X).

The modulus of Z, or $|Z|$, is by definition the distance from the origin O, so that

$$|Z| = \sqrt{(R^2 + X^2)}.$$

The argument of Z is

$$\theta = \arg Z = \tan^{-1} \frac{\text{Im } Z}{\text{Re } Z} = \tan^{-1} \frac{X}{R}.$$

For the series LCR circuit of Fig. 6.14,

$$|Z| = \sqrt{[R^2 + (\omega L - 1/\omega C)^2]},$$

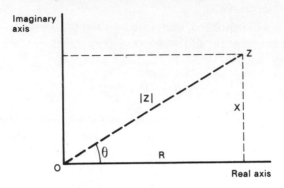

Fig. 6.15 Complex impedance plotted in Argand diagram

and

$$\theta = \tan^{-1} \frac{\omega L - 1/\omega C}{R}.$$

The sign of θ depends on the relative magnitudes of ωL and $1/\omega C$, that is, on whether the circuit is predominantly inductive or capacitive in character.

6.10   Parallel LCR circuit

Suppose that each of the components R, L and C in Fig. 6.16 is pure. This is, of course, an unrealistic assumption, especially for the inductor, which is likely to have appreciable series resistance. Kirchhoff's first law gives

$$i = i_R + i_L + i_C.$$

Here the currents are instantaneous values. With the applied voltage represented as $v_0 \exp j\omega t$, each current will be complex, and will contain a

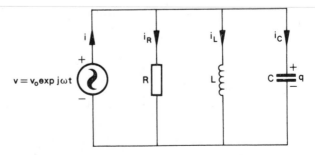

Fig. 6.16 Parallel LCR circuit

factor exp $j\omega t$. Differentiation of a current with respect to time t causes it to be multiplied by the factor $j\omega$. Integration with respect to t causes it to be divided by the factor $j\omega$.

We have

$$v = i_R R,$$

and also

$$v = L\frac{di_L}{dt} = j\omega L i_L.$$

Further

$$v = \frac{q}{C} = \frac{1}{C}\int i_C\, dt = \frac{i_C}{j\omega C}.$$

The original equation for the current can now be written

$$i = \frac{v}{Z} = \frac{v}{R} + \frac{v}{j\omega L} + j\omega C v,$$

or

$$\frac{1}{Z} = \frac{1}{R} + \frac{1}{j\omega L} + j\omega C.$$

Here Z is the effective complex impedance of the combination. The right-hand side of the equation is the sum of the reciprocals of the complex impedances of each branch. In general

$$\frac{1}{Z} = \sum_1^n \frac{1}{Z_j},$$

for complex impedances $Z_1, Z_2, \ldots, Z_j, \ldots, Z_n$ connected in parallel. This is evidently an extension of the rule for conductances in parallel.

6.10.1   Complex admittance

The reciprocal of complex impedance is called the *complex admittance*, and is represented by the symbol Y. Complex admittance is the more convenient quantity for components connected in parallel.

From the general relation of section 6.10 for complex impedances in parallel, we have immediately

$$Y = \sum_1^n Y_j,$$

for the effective complex admittance of the complex admittances Y_1, $Y_2, \ldots, Y_j, \ldots, Y_n$ connected in parallel.

6.10.2   Conductance and susceptance

Let G and B be the real and imaginary parts of the complex admittance Y, so that

$$Y = G + jB.$$

G and B are called the *conductance* and *susceptance* respectively. Complex admittance can be represented as a point in the Argand diagram, with cartesian coordinates (G, B), as in Fig. 6.17.

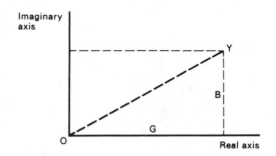

Fig. 6.17 Complex admittance diagram

For a passive circuit component, both the resistance and conductance are necessarily positive. The relation

$$Y = \frac{1}{Z}$$

requires that the reactance and susceptance should be opposite in sign. Further,

$$|Y| = \frac{1}{|Z|},$$

so that

$$\frac{i_{rms}}{v_{rms}} = |Y|.$$

$|Y|$ will be called simply the *admittance*. Also

$$\arg Y = -\arg Z,$$

so that $\arg Y$ is the *lag* of voltage relative to current.

6.11   Imperfect capacitor

There is no loss mechanism in the hypothetical pure capacitor, and the phase angle is therefore $\pi/2$. All practical capacitors exhibit some loss, however, although at moderate frequencies the effect is usually not significant. A small conduction current occurs in electrolytic condensers, where the loss mechanism is best represented by a large shunt resistance. At high frequencies some dissipation occurs in capacitor dielectrics generally, and can be represented by either a large shunt resistance or a small series resistance.

The above effects in combination can be represented by a single shunt resistance, or a single series resistance. In either case the value of the resistance is likely to be frequency-dependent.

Series inductance may also be present, especially in electrolytic condensers. The effect becomes important only at high frequencies, and is non-dissipative. No account will be taken of it here.

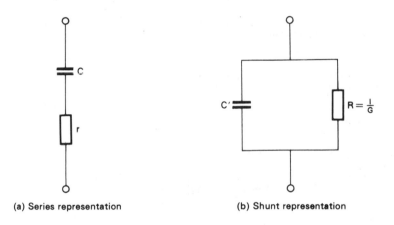

(a) Series representation   (b) Shunt representation

Fig. 6.18 Equivalent circuits for imperfect capacitor (*a*) Series representation, (*b*) shunt representation

The power factor of a capacitor at a given operating frequency is a useful indication of quality. If the effective series resistance is r, or alternatively the effective shunt conductance is G (Fig. 6.18), then the complex impedance and admittance are given respectively by

$$Z = r + 1/j\omega C$$

and

$$Y = G + j\omega C'.$$

Being involved in differing representations of the capacitor, C and C' are not exactly equal.

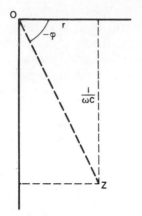
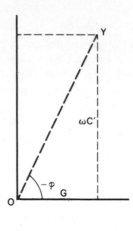

(a) Complex impedance diagram   (b) Complex admittance diagram

Fig. 6.19 Argand diagrams for imperfect capacitor (*a*) Complex impedance diagram, (*b*) complex admittance diagram

In Fig. 6.19, ϕ is the phase lead of voltage over current and is negative for a capacitor. The power factor is therefore

$$\cos \phi = \cos (-\phi) = \frac{r}{|Z|} = \frac{G}{|Y|} .$$

For a capacitor of good quality

$$r \ll \frac{1}{\omega C}, \quad \text{and} \quad G \ll \omega C',$$

so that

$$|Z| = \frac{1}{\omega C}, \quad \text{and} \quad |Y| = \omega C', \text{approximately.}$$

Thus the following approximate relations hold:

Power factor = $\omega C r$, for series representation.

Power factor = $G/\omega C' = 1/\omega C' R$, for parallel representation.

For a good capacitor R is large, and r is small, so that the power factor is not far removed from zero.

6.12   Power matching

For a dc source of emf possessing internal resistance, maximum power can be delivered into a load if this is of equal resistance (section 3.6). The

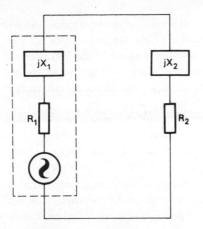

Fig. 6.20 Complex load connected to ac source

internal impedance of an ac source may however be complex. In Fig. 6.20 the source impedance is represented as

$$Z_1 = R_1 + jX_1,$$

and the load impedance as

$$Z_2 = R_2 + jX_2.$$

With a given generator, only R_2 and X_2 can be varied. For maximum power to be dissipated in R_2 maximum current must flow, assuming for the moment the value of R_2 is fixed. Therefore X_2 is adjusted to reduce the total reactance of the circuit to zero, in which case

$$X_2 = -X_1.$$

The reactances are then without effect, and we are back to the familiar situation in which generator and load have pure resistances. For maximum power in R_2, we therefore impose the further condition

$$R_2 = R_1.$$

Both requirements are contained in the relation

$$Z_2 = Z_1^*$$

where Z_1^* is the complex conjugate of Z_1. The condition is said to be one of *conjugate match*. It contrasts with the requirement of *equal* or *identical match* for the matching load of a transmission line (section 7.2.8).

6.13   Worked examples

In the steady state, and for a given source frequency, relations between voltages and currents in linear networks depend only on the circuit components, and are unaffected by the magnitude of the source voltage. In circuit analysis one is concerned with the estimation of impedances and admittances. The process is largely an exercise in complex algebra, and it is usually only at the final stage that direct attention is given to the magnitudes and phases of voltages and currents.

As illustrations we shall examine the behaviour of several simple circuits, each of which exhibits a particular feature of interest.

(*a*) With suitable choice of component values, the total current drawn by the arrangement of Fig. 6.21 is in phase with the supply voltage at all frequencies. The combination then behaves like a pure resistor.

For a branched arrangement like this, it is better to analyse the behaviour in terms of the complex admittance, rather than the complex impedance. The complex admittance is

$$Y = \frac{1}{R_L + j\omega L} + \frac{1}{R_C + 1/j\omega C}.$$

We shall require Y to be real, so that the imaginary part of Y is to be zero.

The temptation to put the terms constituting Y over a common denominator before extracting the imaginary part must be resisted, as this complicates the algebra unnecessarily. Even quite simple ac circuits can be difficult to analyse if the method of solution is unwisely chosen. In the case under consideration treat the two terms separately, multiplying the

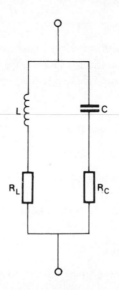

Fig. 6.21

numerator and denominator of each by the complex conjugate of the denominator. It is then easy to extract the imaginary part of Y as

$$-\frac{\omega L}{R_L^2 + \omega^2 L^2} + \frac{1/\omega C}{R_C^2 + 1/\omega^2 C^2}.$$

Setting this equal to zero, rearranging, and inverting each term, one obtains

$$\frac{R_L^2 + \omega^2 L^2}{\omega L} = \frac{R_C^2 + 1/\omega^2 C^2}{1/\omega C}$$

or

$$\frac{R_L^2}{L} + \omega^2 L = \omega^2 C R_C^2 + \frac{1}{C}.$$

This will be satisfied for all ω if the constant terms and the terms in ω^2 are separately equal. Then

$$R_L = R_C = \sqrt{\left(\frac{L}{C}\right)} = R \quad \text{(say)}.$$

The residual admittance reduces to

$$Y = \frac{R}{R^2 + \omega^2 L^2} + \frac{R}{R^2 + 1/\omega^2 C^2},$$

and in view of the above relations, this can be written

$$Y = \frac{R}{R^2 + \omega^2 L^2} + \frac{R}{R^2 + R^4/\omega^2 L^2}$$

$$= \frac{1}{R}.$$

Thus the rather surprising result is obtained that the circuit behaves at all frequencies simply as a fixed pure resistance R. The implication is that the current in each branch can be resolved into two components, one being in phase with the supply voltage and the other in quadrature with it. The components which are in quadrature with the supply voltage have equal amplitudes and are in antiphase with each other. What is the situation with a dc supply?

(*b*) If the resistor R_C is removed, as in Fig. 6.22, the impedance can no longer be real at all frequencies. It is however possible to arrange that at fixed frequency the amplitude of the total current drawn from a constant-voltage source is independent of R.

The admittance is

$$Y = j\omega C + \frac{1}{R + j\omega L}.$$

As the current amplitude is to be constant for a constant supply voltage, $|Y|$ is independent of R. The same must be true of $|Y|^2$.

In investigating this condition it is pointless to put the two terms of the right-hand side of the equation over a common denominator. Proceed instead as follows:

$$|Y|^2 = \left| j\omega C + \frac{R - j\omega L}{R^2 + \omega^2 L^2} \right|^2$$

$$= \left[\frac{R}{R^2 + \omega^2 L^2} \right]^2 + \left[\omega C - \frac{\omega L}{R^2 + \omega^2 L^2} \right]^2.$$

This reduces to

$$|Y|^2 = \omega^2 C^2 + (1 - 2\omega^2 L C)/(R^2 + \omega^2 L^2).$$

This will be independent of R provided

$$2\omega^2 LC = 1,$$

in which event

$$|Y| = \omega C.$$

Although the amplitude of the current is invariant with R, the phase does depend on R. This can be seen by examining

$$\arg Y = \tan^{-1} \frac{\mathrm{Im}\ Y}{\mathrm{Re}\ Y}$$

$$= \tan^{-1} \left[\omega C - \frac{\omega L}{R^2 + \omega^2 L^2} \right] \Big/ \left[\frac{R}{R^2 + \omega^2 L^2} \right].$$

With the aid of the condition

$$2\omega^2 LC = 1,$$

this reduces to

$$\arg Y = \tan^{-1} \left[\frac{R^2 + \omega^2 L^2}{2\omega L} - \omega L \right] \Big/ R$$

$$= \tan^{-1} \left[\frac{R^2 - \omega^2 L^2}{2\omega L R} \right].$$

If R is made very large, $\arg Y \to \tan^{-1} (+\infty) = \pi/2$. This is the lead of current over voltage. This result is to be expected, for then only the

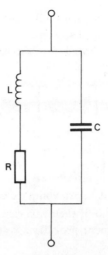

Fig. 6.22

capacitor can influence the phase. A little more care is required in deciding what happens at the other extreme. When R vanishes, the circuit reduces simply to L and C in parallel. The phase of the total current is conferred by the component of greater admittance. Evidently the inductor is dominant, since

$$\omega C = \frac{1}{2} \cdot \frac{1}{\omega L}.$$

Therefore at this extreme the voltage leads the current by $\pi/2$.

An additional deduction is that an intermediate 'in phase' condition exists. Inspection of the expression for arg Y indicates that this occurs for

$$R = \omega L.$$

The circuit then behaves as a pure resistor. To obtain the resistance, it is not necessary to return to the original expression for the admittance Y, because in this special situation

$$Y = |Y| = \omega C = \frac{1}{2\omega L} = \frac{1}{2R},$$

so that the magnitude of the resistance is identified as $2R$.

Some of the results obtained here may seem to contrast strangely with the well-known properties of the so-called *parallel-tuned circuit*. But we shall see later that these are modified by resistive damping, and in the circuit just considered R has been allowed to range from zero to infinity.

(*c*) In the circuit of Fig. 6.23, the low-impedance source provides a voltage v_i of constant frequency.

We shall investigate the amplitude and phase of the output voltage v_o. The resistances R are ganged and equal.

Each branch can be treated as a potential divider connected across the source. The output voltage is

$$v_o = v_i \frac{1}{j\omega C} \Big/ \left[R + \frac{1}{j\omega C} \right] - v_i R \Big/ \left[R + \frac{1}{j\omega C} \right].$$

Therefore

$$\frac{v_i}{v_o} = - \left[R + \frac{1}{j\omega C} \right] \Big/ \left[R - \frac{1}{j\omega C} \right].$$

The quantities v_i and v_o are complex voltages, and can be treated as complex numbers. The ratio of their rms values is $|v_i/v_o|$. This is obviously unity, so that the arrangement provides a voltage gain of unity. The phase retardation of the voltage is

$$\arg \frac{v_i}{v_o} = \arg (-1) + \arg \left[R + \frac{1}{j\omega C} \right] - \arg \left[R - \frac{1}{j\omega C} \right]$$

$$= \pi + \tan^{-1} \frac{-1}{\omega CR} - \tan^{-1} \frac{1}{\omega CR}$$

$$= \pi - 2 \tan^{-1} \frac{1}{\omega CR}.$$

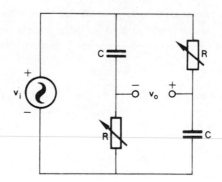

Fig. 6.23

When R is large the output and input voltages are in antiphase. What is then the effective form of the circuit? For $R \to 0$, $\tan^{-1} (1/\omega CR) \to \pi/2$, and the output voltage is in phase with the input voltage. By adjustment of R, continuous variation of phase shift over the range 0 to π is available.

A disadvantage of the circuit is the impossibility of providing a common earth connection for source and load. In addition, the following stage must be of high input impedance, or it will load the phase-shift network and invalidate the above analysis.

Notice that although single reactive components cannot produce a phase change in excess of $\pi/2$, combinations of reactive components can produce larger shifts by their cumulative effect.

The circuit contains resistance, and energy dissipation is therefore an inherent part of its action. Yet the output and input voltages are equal. Is there any conflict between these two statements?

Examples

1. A perfect rectifier with a linear conduction characteristic of slope 1 mA V^{-1} is connected in series with a 1 kΩ ohmic resistor. If the combination is connected to a low impedance 20 V ac supply, calculate the total charge which circulates in
   (a) one period of the 50 Hz alternation, and
   (b) 1000 s.

2. Give values for the rms, peak and peak-to-peak currents in a 120 Ω resistor connected to a low-impedance 240 V ac supply.

3. What are the normal operating resistance and current for a 3 kW heating element designed for a 150 V ac mains supply?

4. The element of a 240 V, 240 W domestic iron is disconnected by its thermostat during half of a particular session of use. In this time what are the mean wattage and rms current?

5. A perfect rectifier with a linear conduction characteristic is connected in a closed series circuit with a 15 Ω resistor, a sinusoidal ac source and a dc moving-coil ammeter. Energy is found to be dissipated in the resistor at the mean rate of 240 W. What is the reading on the ammeter?

6. A 200 V ac supply of low impedance is connected across a 200 Ω resistor. Give values for the wattage, rms current, peak current and mean current.

7. A rectangular voltage waveform has on—off ratio 1:3 and rms value 60 V. What are the peak and mean voltages?

8. A low-impedance 100 V ac supply is connected across 6 kΩ and

4 kΩ resistors in series. Give values for the current and the wattage in the 4 kΩ resistor.

9. An ac current of 120 mA flows through a 100 Ω and a 50 Ω resistor connected in parallel. What current flows in the latter, and what is the total power for the two resistors?

10. What current flows in a 10 μF capacitor connected to a low-impedance 60 V ac supply of angular frequency 300 rad s^{-1}?

11. A 1 μF capacitor is connected to a low-impedance 10 V ac supply of angular frequency 300 rad s^{-1}. Give values for the susceptance of the capacitor and the rms current.

12. What current flows in a 200 mH inductor connected to a low-impedance 60 V ac supply, the angular frequency being 300 rad s^{-1}?

13. A consumer connected to a 200 V public ac electricity supply draws a current of 10 A continuously for 20 hours. If the power factor of his equipment is unity, what is the cost of the electricity used, if the price is 1p per unit (ie per kW hour)?
    If alternatively there is a phase difference of 60° between the supply voltage and the current of 10 A which he draws, what are
    (*a*) his electricity bill, and
    (*b*) the effective or 'in-phase' current for which he pays.
    It is to be assumed that the consumer's electricity meter measures the 'true' wattage.

14. An imperfect 10 μF capacitor is connected across a low-impedance 300 V ac supply, the angular frequency being 300 rad s^{-1}. The capacitor behaves as if shunted by a 3 MΩ resistor. Give values for
    (*a*) the wattless current,
    (*b*) the in-phase current, and
    (*c*) the power consumed.

15. In a series LCR circuit $L = 0.1$ H, $C = 0.1$ μF and $R = 1$ kΩ. For what angular frequency are the current and source voltage in phase? If the source provides 1 V, what are then the voltages across each of the three components?

16. A pure 1 H self inductor is shunted by a 1 kΩ resistor. For an angular frequency of 1000 rad s^{-1} what are the complex admittance,

the admittance, the conductance and the susceptance of the arrangement? Give values for these same quantities when the inductor is replaced by a 1 μF capacitor.

17. A 200 pF capacitor is connected in series with an inductor L. The complex impedance of the arrangement is real for $\omega = 1.3 \times 10^6$ rad s^{-1}. When a resistor R is connected in parallel with the capacitor, the complex impedance is real for $\omega = 1.2 \times 10^6$ rad s^{-1}. Obtain values for L and R.

18. A 200 V 100 W lamp is operated from a low-impedance 250 V ac supply of angular frequency 300 rad s^{-1}, using a series inductor L to drop the voltage to the correct value. A capacitor C is connected directly across the supply and in parallel with the above arrangement in order to restore the overall power factor to unity. Give values for L and C.

19. A capacitor is connected in series with a pure inductor L. The capacitor behaves as a capacitance C shunted by a resistance R. The combination as a whole behaves as a pure resistor at a particular angular frequency ω_0, given by $\omega_0 CR = 2.4$.
(a) For angular frequency ω_0, and if $R = 1.69$ MΩ, what is the effective series resistance of the circuit, and what current is drawn from a low impedance 25 V source?
(b) If R could be made infinitely large, at what angular frequency (in terms of ω_0 only) would the circuit now behave as a pure resistance?

20. The measured self inductance of a coil is 10 mH at an angular frequency of 2000 rad s^{-1}, and 12.8 mH at an angular frequency of 5000 rad s^{-1}. The cause of the variation may be supposed to be shunt capacitance, and resistance effects can be ignored. Give a value for the shunt capacitance. This has been made absurdly large in order to simplify the arithmetic.

21. An ac source provides an output of constant angular frequency 10 000 rad s^{-1}, and behaves like a 1 V source with series resistance 500 Ω and series inductance 5 mH. What simple series network would extract maximum power, and what would be the magnitude of this power?

22. A 100 mH inductor having a series resistance of 100 Ω is suddenly connected across the terminals of a low impedance ac source of

angular frequency 1000 rad s^{-1}. Determine the value of the current flowing 3.142 ms after closure of the circuit, the supply voltage at the instant of connection being the peak value of 100 V. (It is necessary to take into account the effect of a transient which is initiated at the moment of closure of the circuit.)

23. A 1 kΩ resistor and a 0.1 μF capacitor are connected in series. A pure 0.2 H inductor is connected in parallel with the combination. For what angular frequency does the arrangement behave as a pure resistance?

24. When an impure inductor and an impure capacitor are connected in series, at a given frequency the voltages developed across them are in quadrature and equal in magnitude. The impedance of the arrangement is 130 Ω, and the power factor is $\dfrac{12}{13}$. Obtain a value for the inductance, the angular frequency being 1000 rad s^{-1}. For each component the impurity may be taken to have the form of series resistance.

25. A pure inductor L is connected in a closed series circuit with a pure capacitor C and a source of constant emf of variable frequency. A pure resistor of conductance G is connected in parallel with C. Show that when ω is varied, maximum voltage appears across C for

$$\omega^2 LC = 1 - \frac{G^2 L}{2C}.$$

If alternatively C is varied, show that the maximum occurs for

$$\omega^2 LC = 1.$$

26. The source voltage in the circuit of the previous question is 3 V, L is 40 μH and C is 20 nF. When ω is varied the maximum voltage which appears across C is 5 V. Show that the angular frequency is 10^6 rad s^{-1}. If C is now varied with all other quantities remaining at their original optimum settings, show that a maximum voltage of about 5.3 V can be developed across C, with $C = 25$ nF.

7

Filters and transmission lines

The behaviour of a resistive ladder network was investigated in an earlier chapter (3.7). The analysis was restricted to dc conditions, but if the components are purely resistive, the characteristics of a ladder are unchanged with transient or ac excitation. All practical resistors contain imperfections, and these cause modifications in performance as the operating frequency is raised.

If purely reactive components are substituted, energy dissipation within the ladder is eliminated, and all the power delivered into the input end is conveyed to the terminating load. At the same time useful frequency-dependent properties may be acquired.

7.1 Filters

A filter is a network which provides frequency-dependent attenuation. For a well-designed filter, signals within a selected pass band are delivered at the output terminals with undiminished amplitude, while outside this band the attenuation increases rapidly.

7.1.1 Matching load

An iterative ladder network is one constructed of repeated identical elements, which can be represented either as T- or π-sections, as illustrated in Figs. 7.1(a) and (b) respectively. It should be noted that the choice made from these alternatives determines the appropriate forms for the input and load circuits. We shall confine our attention here to T-sections, although the conclusions which will be drawn will be broadly applicable for π-sections.

When the correct matching load is attached to a purely resistive ladder, the input impedance is equal to this load and is independent of the number of sections (3.7.1). We shall see that a similar situation exists for a

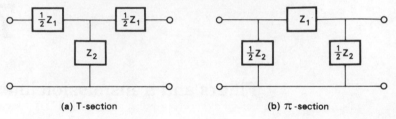

(a) T-section   (b) π -section

Fig. 7.1 Filter circuit elements

ladder constructed of complex impedances, although the required matching load will not now be necessarily real, and is moreover frequency-
dependent even when it is real.

It is helpful to represent the component values of the T-section as in
Fig. 7.2. Here rZ_2 replaces $\frac{1}{2}Z_1$, the dimensionless factor r being in general
complex. A matching load Z_L is connected across the output terminals.
Then we shall require that the input impedance be equal to Z_L, so that

$$Z_L = rZ_2 + 1 \left/ \left(\frac{1}{Z_2} + \frac{1}{rZ_2 + Z_L} \right) \right. .$$

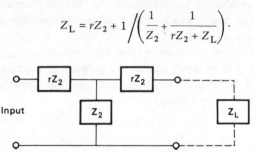

Input

Fig. 7.2 Representative T-section with matching load

This can be very easily reduced to the simple form

$$Z_L^2 = r(r + 2)Z_2^2 .$$

Z_L could in general be complex. It is frequency-dependent unless both Z_2
and r are real.

If π-section representation is used instead, a different matching load
is required, because of the modified configuration of the ladder at each of
its ends.

7.1.2   Propagation constant for matched ladder

The ratio of the input and output voltages for the matched T-section of
Fig. 7.3 is

$$\frac{v_i}{v_o} = \frac{v_i}{\text{voltage across } Z_2} \cdot \frac{\text{voltage across } Z_2}{v_o}$$

If Z is the net complex impedance developed across the terminals of the shunt Z_2, then

$$\frac{v_i}{v_o} = \frac{rZ_2 + Z}{Z} \cdot \frac{rZ_2 + Z_L}{Z_L}$$

$$= \left(1 + \frac{rZ_2}{Z}\right)\left(\frac{rZ_2 + Z_L}{Z_L}\right)$$

$$= \left[1 + rZ_2\left(\frac{1}{Z_2} + \frac{1}{rZ_2 + Z_L}\right)\right]\left(\frac{rZ_2 + Z_L}{Z_L}\right).$$

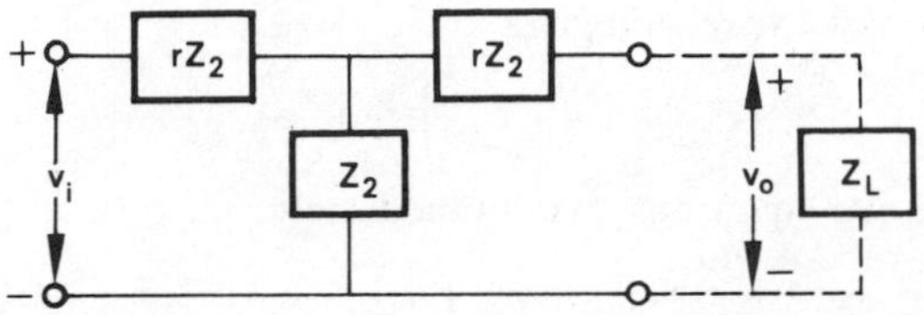

Fig. 7.3 Matched T-section with input and output voltages indicated

This simplifies to

$$\frac{v_i}{v_o} = 1 + r + r(r + 2)\frac{Z_2}{Z_L}.$$

It is convenient to represent this ratio as e^γ, where γ is dimensionless and will in general be complex. γ is sometimes called the propagation constant for the ladder.

7.1.3 Matched lossless ladder

The constituent impedances of a lossless ladder are necessarily pure reactances, so that r is real. It is advantageous in general if the correct matching load Z_L is purely real.

Ladder networks in which the shunt and series reactances are of similar sign prove to be without merit in the context of filter design, and will not be considered here. If the reactances are of unlike sign, then r is negative. The formula which we derived for the matching load (section 7.1.1) gives

$$Z_L = \pm Z_2\sqrt{[r(r + 2)]}.$$

If Z_L is to be real, and since Z_2 is purely imaginary, then $\sqrt{[r(r + 2)]}$ must be purely imaginary. One selects the positive value of the right-hand side of the equation, since Z_L is required to be positive. We can set

$$r(r + 2) = -k^2,$$

where k is real. The solution of this quadratic relation in r is

$$r = -1 \pm \sqrt{(1 - k^2)}.$$

For r real, $k^2 \leqslant 1$, so that k^2 is restricted to the range $0 \to 1$. The permissible range for r is therefore seen to be $-2 \leqslant r \leqslant 0$. If r lies outside this range, then the T-section cannot be matched by a purely resistive load.

For the matched T-section we have in consequence

$$\frac{v_i}{v_o} = e^{\gamma} = 1 + r \pm \sqrt{[r(r + 2)]}.$$

Inversion of this last expression gives

$$e^{-\gamma} = 1 + r \mp \sqrt{[r(r + 2)]}.$$

Addition of the two equations gives immediately

$$\cosh \gamma = 1 + r.$$

Put

$$\gamma = \alpha + j\beta,$$

where α and β are both real. Then

$$\cosh \alpha \cosh j\beta + \sinh \alpha \sinh j\beta = 1 + r,$$

or

$$\cosh \alpha \cos \beta + j \sinh \alpha \sin \beta = 1 + r.$$

The second term on the left-hand side of this equation is equal to zero, as all the other terms are real. The following alternative conclusions may in consequence be drawn:

(*a*) $\sinh \alpha = 0$, so that $\alpha = 0$ and $\cosh \alpha = 1$. Therefore $\cos \beta = 1 + r$, in which case $-2 \leqslant r \leqslant 0$. This has already been seen to be a required condition for the correct matching load to be purely real. With $\alpha = 0$, then $|e^{\gamma}| = |e^{j\beta}| = 1$, and the input and output voltages are equal in amplitude, although differing in phase by an amount depending on r.

(*b*) $\sin \beta = 0$, $\beta = 0$, and $\cos \beta = 1$. Therefore $\cosh \alpha = 1 + r$, in which case $r \geqslant 0$. Then $e^{\gamma} = e^{\alpha}$, and attenuation occurs. The input and output voltages are in phase.

(*c*) $\sin \beta = 0$, $\beta = \pi$, and $\cos \beta = -1$. Therefore $\cosh \alpha = -(1 + r)$, and $r \leqslant -2$. Then $e^{\gamma} = -e^{\alpha}$, and again there is attenuation. The input and output voltages are now in antiphase.

We shall discover in the next section that these conclusions must be applied with some caution.

7.1.4　Filters

The frequency range over which signals are transmitted without attenuation is called the *pass band,* and its limits are defined by condition (*a*):
$-2 \leqslant r \leqslant 0.$

For the T-section of Fig. 7.4(*a*),

$$r = \frac{j\omega L}{1/j\omega C} = -\omega^2 LC,$$

so that the limits of the pass band are determined by

$$-2 \leqslant -\omega^2 LC \leqslant 0.$$

Thus within the band, ω ranges from 0 to $\sqrt{\left(\dfrac{2}{LC}\right)}$. The arrangement constitutes a low-pass filter.

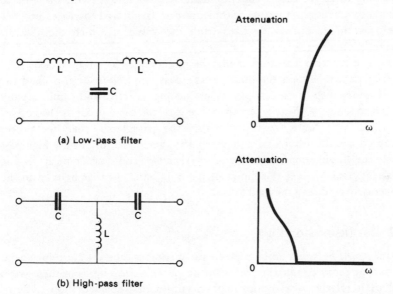

(a) Low-pass filter

(b) High-pass filter

Fig. 7.4 Lossless filter elements (Redrawn from Duffin, W. J. (1965), *Electricity and Magnetism,* McGraw-Hill Book Company.)

For the T-section of Fig. 7.4(*b*),

$$r = \frac{1/j\omega C}{j\omega L} = -\frac{1}{\omega^2 LC},$$

and the limits of the pass band are set by

$$-2 \leqslant -\frac{1}{\omega^2 LC} \leqslant 0.$$

Within the pass band, ω ranges from $\dfrac{1}{\sqrt{(2LC)}}$ to infinity, and the arrangement constitutes a high-pass filter.

We have seen that if the filter is to be matched by a real load, then the value of r is restricted by the condition $-2 \leqslant r \leqslant 0$, which is the range covered by the pass band. Unfortunately, for simple T-sections of the type considered the appropriate value of resistance required for the matching load is itself frequency-dependent, as would be seen by substituting the relations $r = -\omega^2 LC$ and $r = -\dfrac{1}{\omega^2 LC}$ into the equation

$$Z_L = \pm Z_2 \sqrt{[r(r+2)]}.$$

The frequency limits which we have derived for the low- and high-pass filters are valid only if the resistance of the load is adjusted to the correct matching value for each frequency under consideration. With a fixed load resistance attenuation occurs within the pass band, and increases progressively as the frequency deviates from the value at which the filter is correctly matched.

The reactive T-section cannot be matched by a purely resistive load at frequencies beyond the limits of the pass band. Outside this band the input impedance of the low-pass filter becomes reactive and climbs rapidly with frequency. If the source impedance is of the same order as the correct matching load for a frequency within the pass band, then the power accepted by the filter and delivered into the load falls. For a high-pass filter the input impedance becomes reactive and rises rapidly as the frequency falls below the limits of the pass band, so that here again the power delivered into the load falls.

7.2   Transmission lines

Connections between units of electrical apparatus take the form in general of two or more conducting wires. In dc systems the lengths of these wires and their relative dispositions might seem on first consideration to be of no consequence, so long as they provide good conducting paths for the currents they are intended to carry. Unwanted signals may however be picked up from neighbouring electrical equipment. For example, the alternating magnetic field of a transformer will induce an emf in a nearby circuit loop. Pairs of connecting leads must therefore be kept close together to minimise the area capable of intercepting magnetic flux. A problem of comparable magnitude is the existence of stray capacitance between connecting leads and adjacent circuits, giving rise to what is known as electrostatic pick-up. The effect is greatly reduced by enclosing the leads in an earthed sheath, an arrangement well suited for balanced feeders, which alternate in potential in antiphase with respect to each other, and symmetrically with respect to earth. In unbalanced systems, on the other hand, one of the leads is likely to be permanently earthed, so that it can conveniently take the form of an enclosing sheath.

7.2.1 Basic considerations

Overhead telephone lines and the grid electrical supply system use spaced parallel lines, with air as the separating dielectric. These are examples of Lecher wires. For three-phase power supplies, three wires are used as a balanced triplet. For low-power systems, very cheap cables are available, consisting of closely spaced twin wires set in rigid plastic.

Coaxial cables are favoured in those low-power unbalanced systems where mechanical flexibility and electrostatic screening are desired. The copper outer conductor is stranded, and the separating dielectric is pliable. Rigid coaxial cylindrical conductors may be used for higher powers, or where easy access is required to the space between the conductors.

The short 'slab' lines which are sometimes used at very high frequencies are constructed as two fairly closely spaced, rigid, conducting, flat strips. The arrangement is usually restricted to unbalanced systems, and for simplicity the earthed strip is often the metal baseplate of the unit.

A complication associated with non-permanent cable connections is the need to terminate cable sections with special connecting plugs and sockets. At high frequencies, a badly designed connection can seriously modify the electrical performance of a cable link. Coaxial cable connectors are costly, and the process of attaching them to the cable is time consuming.

As the operating frequency is raised, an unscreened line behaves as an increasingly efficient aerial, and wastefully radiates energy. In addition, the dielectric spacing material is itself a source of energy dissipation at high frequencies. At the present time transmission lines are therefore unsuited for long-distance, high-frequency links.

7.2.2 Representative line element

At low frequencies, a short length of transmission line which has no load connected behaves as a 'lumped' capacitor.

The magnetic field distribution associated with the flow of current in the two conductors gives rise to series inductance. The effect is not wholly confined to one or other of the conductors, but it is convenient in circuit analysis to represent it as affecting only one of them.

In a uniform transmission line, capacitance and inductance are distributed effects, so that currents and voltages on the line are also continuously distributed. If a realistic analysis of behaviour is to be made, the representative line element must therefore be of infinitesimal length. T- or π-section configurations can be used, the former being illustrated in Fig. 7.5.

Notice that L and C have the dimensions here of inductance per unit length and capacitance per unit length respectively.

Some ohmic resistance is present in the metal conductors of the line. This increases in magnitude as the frequency rises, because current flow is then restricted by skin effect to the surfaces of the conductors, so that the effective cross-section is reduced. The effect can be represented as series resistance in the line element.

With modern dielectric materials, resistive leakage between conductors is quite negligible at low frequencies. At high frequencies, the rapidly alternating electric field gives rise to significant dielectric loss. The effect is best depicted in the line element as shunt conductance.

Analysis of transmission line behaviour is, of course, more difficult if allowance has to be made for loss. The more important electrical characteristics of transmission lines are fortunately not greatly influenced by distributed loss-mechanisms, and we shall therefore restrict our attention to the simple lossless model of Fig. 7.5.

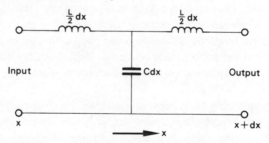

Fig. 7.5 Representative element of transmission line

7.2.3   Characteristic impedance

The limits of the pass band of a matched T-section are given (section 7.1.4) by the condition

$$-2 \leqslant r \leqslant 0.$$

For the lossless transmission line element of Fig. 7.5

$$r = \frac{Z_1}{2Z_2} = -\frac{LC(\omega \, dx)^2}{2},$$

so that the extent of the pass band is determined by the condition

$$-2 \leqslant -LC \left(\frac{\omega \, dx}{2} \right)^2 \leqslant 0.$$

The pass band is therefore unlimited, since dx is vanishingly small.

The matching load is given (section 7.1.1) by the relation

$$Z_L^2 = r(r + 2)Z_2^2 = -\frac{LC(\omega \, dx)^2}{2} \cdot 2 \cdot \left(\frac{1}{j\omega C \, dx} \right)^2,$$

where r has been neglected in comparison with the finite number 2. Simplifying,

$$Z_L = \sqrt{\frac{L}{C}}.$$

The matching load of a transmission line is called the characteristic impedance, and is more usually represented by the symbol Z_0. It is evidently real, positive, and independent of frequency. When a line is terminated by the correct matching load, the input resistance is equal to Z_0 at all frequencies, and is unaffected by the length of the line. Electrical energy supplied at the input end is conveyed along the line to the matching load without loss. In many respects the arrangement behaves as if the transmission line were absent, the load being connected directly to the terminals of the source.

7.2.4 Equations of telegraphy

The behaviour of the lossless transmission line can be more easily investigated in terms of the representative element illustrated in Fig. 7.6. A single inductor $L dx$ is shown, in place of the two inductors $\frac{1}{2}L dx$ depicted earlier. Potential differences and currents at the input and output connections of the element are indicated.

The current through the capacitor is the product of the capacitance and the time rate of change of the applied potential difference. Applying Kirchhoff's first law to the junction at the upper connection of the capacitor, we have

current arriving = sum of currents leaving.

Thus

$$i = \left(i + \frac{\partial i}{\partial x} \, dx \right) + C dx \, \frac{\partial}{\partial t} \left(v + \frac{\partial v}{\partial x} \, dx \right).$$

Ignoring the last term, which is of the second order of smallness, and simplifying,

$$\frac{\partial i}{\partial x} = -C\,\frac{\partial v}{\partial t}\,.$$

This is the First Equation of Telegraphy.

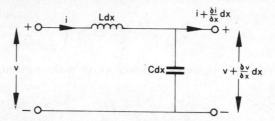

Fig. 7.6 Alternative representation of transmission line element

Applying Kirchhoff's second law to a path traced clockwise round the element and passing through the four terminals:

$$L\,dx\,\frac{\partial i}{\partial t} + \left(v + \frac{\partial v}{\partial x}\,dx\right) - v = 0,$$

or

$$\frac{\partial v}{\partial x} = -L\,\frac{\partial i}{\partial t}\,.$$

This is the Second Equation of Telegraphy.

7.2.5   Phase-velocity

If the First and Second Equations of Telegraphy are differentiated partially with respect to t and x respectively, then the following relations can be extracted:

$$\frac{\partial^2 i}{\partial x\,\partial t} = -C\,\frac{\partial^2 v}{\partial t^2} = -\frac{1}{L}\,\frac{\partial^2 v}{\partial x^2}\,.$$

Rearranging,

$$\frac{\partial^2 v}{\partial x^2} - LC\,\frac{\partial^2 v}{\partial t^2} = 0.$$

Let

$$c^{*2} = \frac{1}{LC}\,.$$

Then

$$\frac{\partial^2 v}{\partial x^2} - \frac{1}{c^{*2}} \frac{\partial^2 v}{\partial t^2} = 0.$$

This is a one-dimensional wave-equation. It is easy to verify by substitution that the solution is

$$v = f(x \mp c^* t),$$

where the form of the function f depends on the form of the driving function at the input end.

Substitute the above solution into the Second Equation of Telegraphy, and integrate the resulting equation with respect to time. Rearrangement gives

$$Lc^* i = \pm f(x \mp c^* t).$$

The quantity Lc^* has the dimensions of impedance, and is equal to $\sqrt{\dfrac{L}{C}}$, the characteristic impedance.

The above solutions indicate that voltage and current are propagated as wave-motions along a transmission line. The location on the line of a given magnitude of potential difference, or of current, moves in such a way that the phase $x \mp c^* t$ remains constant. This requires

$$d(x \mp c^* t) = 0$$

or

$$dx \mp c^* dt = 0,$$

giving

$$\left(\frac{dx}{dt}\right)_{\substack{\text{constant} \\ \text{phase}}} = \pm c^*.$$

The quantity on the left-hand side of this equation is called the *phase-velocity*, and it is positive if the phase is represented as $x - c^* t$, implying forward motion in the direction of increasing x. A negative value is associated with a backward-travelling wave, which travels towards the input end.

The occurrence of wave-motion on the line means that changes at one end cannot produce simultaneous changes at the other. This situation contrasts with a purely resistive ladder network in which all current paths are made very short, where a change at the input produces its full effect almost instantly everywhere on the ladder.

The phase-velocity on a lossless line is independent of frequency, so that all the components of differing frequencies constituting an impulse travel at the same speed, and the shape of the impulse remains constant. In real transmission lines there is always some loss. This gives rise to frequency-dependence, or *dispersion* as it is called, of phase-velocity, so that complex signals experience phase-distortion. The effect can be serious in lines used for long-distance communication, and artificial correction may be essential.

7.2.6   Line constants

For a cylindrical coaxial line, the capacitance per unit length is

$$C = 2\pi\epsilon\epsilon_0/\log_e(b/a) \text{ F m}^{-1},$$

where a is the outer radius of the inner conductor, b is the inner radius of the outer conductor, and

$$\epsilon_0 = \frac{10^7}{4\pi c^2} \text{ F m}^{-1}.$$

c being the speed of light in a vacuum in m s^{-1}. ϵ is the dielectric constant of the material filling the interspace.

The inductance per unit length, L, depends on frequency. For low frequencies the electric currents are distributed uniformly over the cross-sections of the conductors, and the magnitude of L depends on, amongst other things, the thickness of the outer conductor. At high frequencies, current flow is restricted to the outer surface of the inner conductor and the inner surface of the outer conductor. In the absence of ferrous material the self inductance per unit length is then simply

$$L = \frac{\mu_0}{2\pi} \log_e (b/a) \text{ H m}^{-1},$$

where

$$\mu_0 = 4\pi \times 10^{-7} \text{ H m}^{-1}.$$

If both conductors are tubular, this formula is also valid at low frequencies, provided the thickness of the conductors is small compared with the mean radii.

Using the above values for C and L, the matching load for a coaxial line is

$$Z_0 = \sqrt{\frac{L}{C}} = \sqrt{\left(\frac{\mu_0}{\epsilon_0\epsilon}\right)} \frac{\log_e(b/a)}{2\pi}$$

$$= \frac{60}{\sqrt{\epsilon}} \log_e (b/a) \quad \text{ohm}.$$

Z_0 is seen to depend only on the ratio of conductor radii, and the dielectric constant. The quantity $\dfrac{\log_e(b/a)}{\sqrt{\epsilon}}$ does not usually differ greatly from unity, and coaxial cables are commonly designed for a matching load of 75 ohm. This gives rise to problems in impedance matching when connection is to be made to circuits of high impedance.

For a pair of well-spaced parallel wires in air,

$$Z_0 = 120 \log_e(d/a) \quad \text{ohm},$$

d being the spacing between centres, and a the radius of each wire. This configuration enables high Z_0 values to be achieved, since it is easy to make d very much greater than a, although at high frequencies there is then the problem of excessive energy loss by radiation.

The phase-velocity on a transmission line is

$$c^* = 1/\sqrt{(LC)},$$

so that for a coaxial cable,

$$c^* = 1/\sqrt{(\mu_0\epsilon_0\epsilon)}.$$

This is noticeably independent of cable geometry. It can be shown to be the same as the phase-velocity of electromagnetic waves travelling in the unbounded dielectric. This is not wholly surprising, for the moving voltage and current distribution on the cable is necessarily accompanied by transverse electric and magnetic fields. If the space between the conductors is occupied by air or vacuum, the phase-velocity is equal to the speed of light in empty space, 3×10^8 m s^{-1}. The same formula for the phase-velocity is obtained for a separated twin-line structure, provided the lateral extent of the supporting dielectric greatly exceeds the separation between centres.

In general, for single-frequency wave-motion,

$$c^* = f\lambda,$$

where f is the frequency and λ the corresponding wavelength. At low frequencies the wavelength on a transmission line is large, amounting to 6000 km at mains frequency on an air-filled line. In the majority of low-frequency applications the line length l is minute in comparison with the wavelength, and there is effectively no variation in voltage or current along the line. At microwave frequencies, on the other hand, several wavelengths may be contained in a given length of connecting cable. For an air-filled line, the wavelength is 0.1 m at a frequency of 3 GHz. In such circumstances, the line is likely to behave as an integral part of the circuits attached at each end, unless it is correctly matched at the output end.

7.2.7   *Current and voltage distributions*

If the signal at the input end of a transmission line is sinusoidal, the time-dependence everywhere on the line can be represented by the factor $\exp j\omega t$. The wave-equation of 7.2.5:

$$\frac{\partial^2 v}{\partial x^2} - \frac{1}{c^{*2}} \frac{\partial^2 v}{\partial t^2} = 0$$

therefore becomes

$$\frac{\partial^2 v}{\partial x^2} + \frac{\omega^2}{c^{*2}} v = 0.$$

Put

$$\beta = \frac{\omega}{c^*}.$$

Then alternative solutions of the wave-equation contain the factors $\exp(j\beta x)$ and $\exp(-j\beta x)$, so that a complete solution which takes into account the time-dependence is

$$v = (A e^{-j\beta x} + B e^{j\beta x}) e^{j\omega t},$$

where A and B are constants. The first and second terms on the right-hand side of this equation are associated with forward- and backward-travelling waves respectively.

Substituting the above solution into the Second Equation of Telegraphy:

$$\frac{\partial v}{\partial x} = -L \frac{\partial i}{\partial t}$$

gives

$$-j\beta(A e^{-j\beta x} - B e^{j\beta x}) e^{j\omega t} = -j\omega L i,$$

or

$$\frac{\omega L}{\beta} i = (A e^{-j\beta x} - B e^{j\beta x}) e^{j\omega t}.$$

Here

$$\frac{\omega L}{\beta} = c^* L = \sqrt{\frac{L}{C}} = Z_0.$$

Thus

$$Z_0 i = (A e^{-j\beta x} - B e^{j\beta x}) e^{j\omega t}.$$

7.2.8   Input impedance

The complex input impedance of the line at any section is

$$Z_i = \frac{v}{i} = \frac{Ae^{-j\beta x} + Be^{j\beta x}}{Ae^{-j\beta x} - Be^{j\beta x}} Z_0.$$

Let the origin of x be located at the terminating load Z_L, as in Fig. 7.7. Then x is actually negative at all points on the line. As $x = 0$ at the load,

$$Z_L = \frac{A + B}{A - B} Z_0,$$

so that

$$A(Z_L - Z_0) = B(Z_L + Z_0).$$

Using this relation, the expression for the input impedance becomes

$$Z_i = \frac{(Z_L + Z_0)e^{-j\beta x} + (Z_L - Z_0)e^{j\beta x}}{(Z_L + Z_0)e^{-j\beta x} - (Z_L - Z_0)e^{j\beta x}} Z_0.$$

On rearranging, this reduces to

$$Z_i = \frac{Z_L \cos \beta x - jZ_0 \sin \beta x}{Z_0 \cos \beta x - jZ_L \sin \beta x} Z_0.$$

For a line of length l, the x coordinate at the input end is $-l$, so that

$$Z_i = \frac{Z_L \cos \beta l + jZ_0 \sin \beta l}{Z_0 \cos \beta l + jZ_L \sin \beta l} Z_0.$$

The input impedance is seen to depend in a rather complicated way on the load, and on the line length. The line transforms the load impedance Z_L to the input impedance Z_i, without itself consuming power, and can be regarded as an impedance transformer.

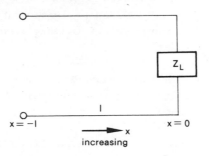

Fig. 7.7 Line terminated with load Z_L

If the load matches the line, $Z_L = Z_0$, and the above relation reduces predictably to $Z_i = Z_0$. It is noticeable, furthermore, that in this case $B = 0$, so that no backward-travelling wave is present on the line.

The input impedance of a mismatched line is frequency-dependent, and in high-power applications the presence of backward wave may give rise to dielectric breakdown, and also to local heating. The restricted range of characteristic impedance in commercially available cables makes precise matching difficult. If power-transference is not a primary consideration, a high impedance load can be matched to a line by shunting it with an appropriate resistor. This is a common procedure, and where it is unacceptable, more elaborate matching techniques have to be employed.

Examples

1. A T-section filter element consists of two identical series capacitors C, and a 50 mH inductive shunt. Obtain a value for C if the filter is to match 50 Ω for $\omega = 1000$ rad s^{-1}.

2. A T-section filter element consists of two identical series inductors L, and a 5 μF capacitive shunt. Obtain a value for L if the filter is to match 50 Ω for $\omega = 4000$ rad s^{-1}.

3. A T-section filter element consists of two identical series capacitors C, and a 20 mH inductive shunt. Obtain a value for C if the filter is to match 100 Ω for $\omega = 5000$ rad s^{-1}. Over what range of angular frequency is the attenuation zero, assuming a correct matching load to be always provided?

4. A T-section filter element consists of two identical series inductors L, and a 10 μF capacitive shunt. Obtain a value for L if the filter is to match 50 Ω for $\omega = 2000$ rad s^{-1}. Up to what maximum angular frequency is the attenuation zero, assuming a correct matching load to be always provided?

5. A T-section filter element consists of two identical series capacitors, and an inductive shunt. What is the appropriate matching load at the lower limit of the pass band?

6. The T-section elements of a ladder-type filter embody 4/3 μF shunting capacitors and 20 mH series inductors. What is the appropriate matching load for an angular frequency of 5000 rad s^{-1}?

7. Give the value of the current in a matched 50 Ω transmission line connected to a 50 mV low-impedance source.

8. The emf of a generator is 10 V, and the series internal resistance is 50 Ω. What is the power supplied to a 50 Ω load which is connected via a 50 Ω transmission line?

9. A 150 Ω load is connected to a 15 mV source by a 75 Ω line. Matching is achieved by connecting a 150 Ω resistor directly across the load. If the power consumed by the load is 375 nW, what is the internal resistance of the source, supposed real?

10. The phase-velocity of electrical disturbances on a given lossless transmission line is half the speed of light *in vacuum*. What is the effective value of the dielectric constant of the medium?

11. Give values for the distributed electrical constants of a 60 Ω air-cored lossless line.

12. Obtain an expression for the input impedance of a uniform lossless transmission line of characteristic impedance Z_0, which is a quarter of a wavelength long at the operating frequency $\left(\beta l = \dfrac{\pi}{2}\right)$ and which is terminated by a load Z_L. Show that if Z_L is real, then the input impedance is also real.

   The series internal resistance of a source is 100 Ω, and it is to be used to supply a 144 Ω load. Suggest a suitable characteristic impedance for a quarter-wave lossless transmission line to be used to connect the source to the load. Discuss whether maximum power is delivered into the load, and whether there is standing wave on the line.

Resonance

8.1 General considerations

Resonance is said to occur when a system capable of free vibration at a certain natural frequency is forced into a maximum amplitude of response by suitable adjustment of the frequency of a source of continuous excitation. The nature of the response depends on the constants of the system, and sometimes resonances of differing kinds can be produced within quite a narrow frequency spectrum. Mechanical systems exhibit responses which can be conveniently designated displacement, velocity and acceleration resonances, and the responses of electrical systems might correspondingly be described as charge, current or voltage resonances. Some care is necessary in describing a particular resonant response, if confusion in identification is to be avoided.

It is sometimes said that at resonance the complex impedance of an electric circuit becomes real, and that for a given amplitude of excitation maximum energy is then dissipated in the system. These statements are valid for the simple LCR series circuit, but are in general misleading and may be at best only approximately correct.

The most valued property of resonant circuits is their selectivity. This can be exploited for the suppression of generator overtones, mains hum and unwanted signals generally. Resonant circuits are provided in tuned amplifiers as loads for the amplifying devices, so that only the intended frequency band is passed. The addition of resonant circuits to filter networks can provide extra stop bands, and enable existing pass bands to be suitably reshaped.

8.2 Series resonance

Consider the series LCR circuit of Fig. 8.1. The source will be supposed to provide an ac voltage of constant amplitude, and to be adjustable in

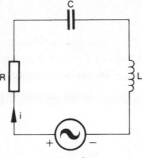

Fig. 8.1 Series LCR circuit $v_0 \exp j\omega t$

frequency. The three components L, C and R are pure and independent of frequency.

We have previously investigated the transient behaviour of the LCR series circuit at some length. With the provision of an ac source we are now concerned with the response of a linear electrical system to continuous sinusoidal forcing. The behaviour of this driven system differs markedly from the transient response of the free-running system.

8.2.1   Resonant frequency

The complex impedance of the circuit is

$$Z = R + j\omega L + \frac{1}{j\omega C}.$$

We shall investigate the condition of maximum current with variation of frequency, or *current resonance* as it could be called. The phase of the current will not be of direct interest, and in view of the constancy of the supply voltage we shall require simply that $|Z|$ be minimised.

Now

$$|Z|^2 = R^2 + \left(\omega L - \frac{1}{\omega C} \right)^2.$$

Since the term in brackets cannot be negative, the minimum value of $|Z|$ is R, and this occurs for a particular angular frequency ω_r given by

$$\omega_r L = \frac{1}{\omega_r C},$$

or more simply

$$\omega_r^2 LC = 1.$$

This relation determines the frequency for resonance.

8.2.2   *Voltage gain*

The current in the circuit varies inversely with $|Z|$. The variation with angular frequency is illustrated in Fig. 8.2. The current is seen to approach zero at the extreme values of frequency. Notice that the curve has finite slope at the origin.

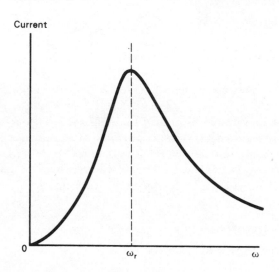

Fig. 8.2 Variation of current with angular frequency for series LCR circuit supplied by constant-voltage source

The impedance at resonance is real and equal to the resistance R. The voltage developed across R is then equal to the voltage provided by the source, and the effect is as if the reactive components were absent. At frequencies above resonance the reactance is positive, and the circuit is predominantly inductive in behaviour. At frequencies below resonance the reactance is negative, and the circuit is capacitive in behaviour.

The relatively large current i which flows at resonance produces large voltages of equal amplitudes across the reactive components. The moduli of these voltages are

$$|v_L| = \omega_r L \, |i|,$$

and

$$|v_C| = |i|/\omega_r C.$$

The voltages are in antiphase, and their sum is therefore zero.

The applied voltage is equal to $R|i|$. The circuit therefore provides a voltage gain equal to

$$\left| \frac{v_\mathrm{L}}{\text{applied voltage}} \right| = \frac{\omega_\mathrm{r} L}{R},$$

or

$$\left| \frac{v_\mathrm{C}}{\text{applied voltage}} \right| = \frac{1}{\omega_\mathrm{r} C R}.$$

This quantity is usually represented by the symbol Q. If R is small, the voltage gain can be quite large, but the resistance presented to the source is then correspondingly small. This passive amplifier is no more than an impedance transformer, and the voltage amplification afforded by the series tuned circuit is merely a consequence of the associated step-up in impedance.

8.2.3   Half-power points

The electric power delivered into a circuit of complex impedance Z by an ac source providing voltage and current of amplitudes v_0 and i_0 respectively is

$$P = \frac{v_0 i_0}{2} \cos \phi = \frac{v_0^2}{2|Z|} \cos \phi,$$

where $\cos \phi$ is the power factor. For the series LCR circuit

$$Z = R + j \left(\omega L - \frac{1}{\omega C} \right),$$

so that

$$\cos \phi = \frac{\mathrm{Re}\, Z}{|Z|} = \frac{R}{|Z|}.$$

Therefore

$$P = \frac{v_0^2 R}{2|Z|^2}.$$

If the source impedance is negligible, v_0 is constant, and the numerator is independent of frequency. At resonance $|Z|$ takes a minimum value equal to R, and the power delivered into the circuit then rises to the maximum value

$$P_{\max} = \frac{v_0^2}{2R}.$$

We shall now set out to discover at what frequencies the power falls to half this value. The required condition is

$$\frac{v_0^2 R}{2|Z|^2} = \frac{P_{max}}{2} = \frac{v_0^2}{4R},$$

so that

$$|Z|^2 = 2R^2.$$

It follows that

$$|Z|^2 = R^2 + \left(\omega L - \frac{1}{\omega C}\right)^2 = 2R^2,$$

or

$$R^2 = \left(\omega L - \frac{1}{\omega C}\right)^2.$$

This relation indicates that at the half-power points the resistance and reactance of the circuit are equal in magnitude. It follows that the phase angle is then $\pm(\pi/4)$, and that the power factor is $1/\sqrt{2}$.

We have on further rearrangement

$$\omega^2 L \pm \omega R - \frac{1}{C} = 0.$$

The solutions of this quadratic in ω are given by

$$2\omega L = \mp R \pm \sqrt{\left(R^2 + \frac{4L}{C}\right)}.$$

Only positive roots are physically meaningful. Let these be ω_1 and ω_2, where

$$2\omega_1 L = -R + \sqrt{\left(R^2 + \frac{4L}{C}\right)},$$

and

$$2\omega_2 L = R + \sqrt{\left(R^2 + \frac{4L}{C}\right)}.$$

Evidently ω_1 and ω_2 are disposed symmetrically about a mean value ω_m given by

$$2\omega_m L = \sqrt{\left(R^2 + \frac{4L}{C}\right)}.$$

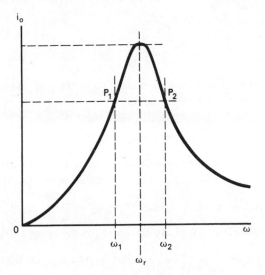

Fig. 8.3 Half-power points for LCR series circuit

If

$$R^2 \ll \frac{4L}{C},$$

then

$$\omega_m^2 LC = 1, \quad \text{approximately},$$

so that ω_m is close to the value ω_r for series resonance.

The difference in frequencies between the roots is given by

$$2(\omega_2 - \omega_1)L = 2R.$$

It follows that

$$\frac{\omega_r}{\omega_2 - \omega_1} = \frac{\omega_r L}{R},$$

which we recognise as the Q of the circuit.

Let us now return briefly to the relation

$$|Z|^2 = 2R^2,$$

which is valid at the half-power points. The amplitudes of voltage and current are related by

$$\frac{v_0}{i_0} = |Z|.$$

It follows that at the half-power points

$$i_0 = \frac{v_0}{\sqrt{2}R}.$$

This is $1/\sqrt{2}$ of the amplitude at resonance. Note the positions of the half-power points P_1 and P_2 in the response curve of Fig. 8.3.

8.2.4   *Quality factor and selectivity*

The sharpness of resonance, or *selectivity* as it is alternatively known, depends on the closeness in frequency of the half-power points. Since the Q, or *quality factor* as it is called, is equal to $\dfrac{\omega_r}{\omega_2 - \omega_1}$, it is a useful measure of the selectivity. We have seen that for the series LCR circuit the Q is equal to $\dfrac{\omega_r L}{R}$ or $\dfrac{1}{\omega_r CR}$. The necessary conditions for high selectivity are evidently high inductance, and low capacitance and ohmic loss.

8.3   Q-meter

The majority of impedance-measuring instruments exploit the bridge principle, in which a state of balance in a network is revealed by the null response of a detecting device. The Q-meter involves a quite different mode of operation, depending for its action on the voltage gain produced by a series LCR circuit.

The instrument is used for the measurement of admittance. The essential features of the measuring circuit are illustrated in Fig. 8.4. The frequency of the ac source is variable over a wide range. The internal impedance is only a fraction of an ohm, so that the output voltage varies little with change of circuit impedance. This voltage is adjustable, and a constant reading v_i can therefore be maintained on the voltmeter connected across the source. The unknown admittance is connected in parallel

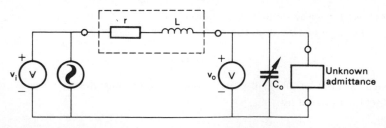

Fig. 8.4 Q-meter measuring circuit

with the variable calibrated capacitor C_0, and various low-loss plug-in coils L are provided for each frequency range of the instrument. At each stage in a series of measurements C_0 is adjusted so that a maximum voltage is recorded by the high-resistance voltmeter connected across it. The state of the circuit then differs from the resonant condition of the series LCR circuit previously discussed (section 8.2), where the current drawn from the source and in consequence the voltage developed across the capacitor were maximised by adjustment of the frequency of the source.

8.3.1 Susceptance measurement

It will now be shown that the Q-meter provides a simple substitution technique for the measurement of susceptance, which may be positive or negative. The analysis which follows has been simplified without loss of generality by representing the unknown admittance as a parallel combination of a capacitance C' and a conductance G (Fig. 8.5). Put

$$C = C' + C_0.$$

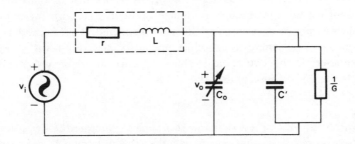

Fig. 8.5 Q-meter circuit with capacitive representation of unknown admittance

Then

$$\frac{v_i}{v_o} = \frac{r + j\omega L + 1/(j\omega C + G)}{1/(j\omega C + G)}$$

$$= 1 + (r + j\omega L)(G + j\omega C)$$

$$= (1 - \omega^2 LC + rG) + j\omega(LG + Cr).$$

Therefore

$$\left|\frac{v_i}{v_o}\right|^2 = (1 - \omega^2 LC + rG)^2 + \omega^2(LG + Cr)^2.$$

The applied voltage is maintained constant. If now $|v_o|$ is maximised by adjustment of C_0, it follows that $\left|\dfrac{v_i}{v_o}\right|^2$ is minimised, so that

$$\frac{\partial}{\partial C_0}\left|\frac{v_i}{v_o}\right|^2 = 0.$$

An equivalent condition is

$$\frac{\partial}{\partial C}\left|\frac{v_i}{v_o}\right|^2 = 0,$$

and therefore

$$-2\omega^2 L(1 - \omega^2 LC + rG) + 2\omega^2 r(LG + Cr) = 0.$$

This simplies to

$$C = L/(r^2 + \omega^2 L^2).$$

The critical value for C is seen to be determined by the constants L and r of the coil, and the angular frequency ω. Notice that it is independent of the conductance G. The unknown susceptance can therefore be measured by a direct substitution technique, involving adjustment of the calibrated capacitor C_0 for a maximum of voltage across it, first with the unknown connected into the circuit and then disconnected. The method is valid whether or not there is conductance associated with the unknown susceptance. If the two critical values of C_0 are successively C_{01} and C_{02}, then the unknown susceptance is

$$\omega C_{02} - \omega C_{01}.$$

C_{02} will be greater or less than C_{01} according to whether the susceptance is positive or negative, that is, on whether the unknown is capacitive or inductive in character.

There is always some stray conductance associated with the calibrated capacitor. This contributes no error if its magnitude is unchanged by the setting of C_0. In practice there is variation with C_0, so that in the preceding calculation the effective value of G changes with C, and the unknown susceptance will not be exactly equal to $\omega C_{02} - \omega C_{01}$. In most practical situations the error is fortunately small.

Provided circuit losses are small, in the vicinity of the maximum the voltage across C_0 varies sharply with adjustment of C_0, and the critical value is well defined.

8.3.2   Conductance measurement

If the voltage applied to the Q-meter measuring circuit is maintained constant at a selected value, the voltage developed across the calibrated capacitor will be proportional to the quality factor of the circuit. The scale of the voltmeter which records this voltage can therefore be calibrated directly in terms of Q values. It will now be shown that its readings can be used for the measurement of conductance.

With the unknown admittance connected, the maximum quality factor is given by

$$\left|\frac{v_i}{v_o}\right|^2 = \frac{1}{Q^2} = (1 - \omega^2 LC + rG)^2 + \omega^2 (GL + Cr)^2.$$

In the absence of conductance let the maximum quality factor have the value Q_0. Then

$$\frac{1}{Q_0^2} = (1 - \omega^2 LC)^2 + \omega^2 C^2 r^2.$$

The condition

$$C = L/(r^2 + \omega^2 L^2),$$

which has already been derived (section 8.3.1), holds whether the unknown is connected into the circuit or not. The constants L and r of the coil are not directly observable, but can be eliminated between the above three equations to give

$$G = \omega C \left(\frac{Q_0}{Q} - 1\right) (Q_0^2 - 1)^{-1/2}.$$

This relation enables a value for G to be determined. The value to be used for C is the critical value with the unknown admittance disconnected. The formula is inexact if there is conductance associated with the calibrated capacitor, whether fixed or dependent on its setting.

8.4 Quality factor for nonresonant elements

Quality factor can be defined by the alternative relation

$$Q = 2\pi \frac{\text{maximum energy stored}}{\text{energy dissipated in one cycle}}.$$

For a coil of inductance L and series resistance R, in which a current $i_0 \cos \omega t$ is flowing, this gives

$$Q = 2\pi \frac{\frac{1}{2}L i_0^2}{\frac{1}{2}R i_0^2 T} = \frac{\omega L}{R},$$

where T is the period of the alternation. For a condenser which has capacitance C and series resistance R, and carries a current of similar form, it is

$$Q = 2\pi \frac{\frac{1}{2}Cv_0^2}{\frac{1}{2}Ri_0^2 T}.$$

Here v_0 is the maximum potential difference, equal to $\dfrac{i_0}{\omega C}$, so that

$$Q = \frac{1}{\omega CR}.$$

Both these results are compatible with the values obtained previously for the series LCR circuit. The definition used here has the advantage of being meaningful both within and outside the context of resonance, and of being applicable to circuit elements which are not in themselves capable of resonance.

8.5   Parallel resonance

A loss-free parallel circuit would consist of a pure inductor connected in parallel with a pure capacitor as in Fig. 8.6. The complex admittance of such an arrangement is

$$j\left(\omega C - \frac{1}{\omega L}\right).$$

This is zero at a particular angular frequency ω_r, given by

$$\omega_r^2 LC = 1,$$

and becomes infinite in magnitude at either extreme of frequency.

It was convenient to discuss the behaviour of the series LCR circuit in terms of the current which it would draw from a constant-voltage

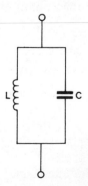

Fig. 8.6 Lossless parallel circuit

source. A constant-current source is more appropriate for the parallel LC circuit. At the resonant frequency an infinite voltage develops across the terminals, the effect being best described as *voltage resonance*. The currents which flow in the two branches of the circuit are of infinite amplitude, and are in precise antiphase with each other. This lossless circuit could therefore produce infinite current gain.

These extreme properties are modified in practice by strays. The least significant are distributed reactances, such as capacitance between the turns of the coil, and series lead inductance of the capacitor. Resistive damping is of much greater importance, electrical energy being converted to heat in the windings of the coil and in the former on which it is wound. Energy is also lost by radiation. At a given frequency these effects can be represented by resistance connected either in series or in parallel with the coil, but for a finite range of frequency a combination of both is required. Imperfections in the dielectric of the capacitor can also contribute both series and shunting resistance.

The behaviour of any model of the parallel circuit which makes allowance for all these effects would be very difficult to analyse. For simplicity we shall deal separately with the influences of shunting and series resistances.

8.5.1   Parallel combination with shunting resistance

Resistances shunting the coil and capacitor of a reactive parallel combination will be represented by a single ohmic conductance G, as in Fig. 8.7. Let the source be again of the constant-current type. The complex admittance of the arrangement is

$$Y = G + j\left(\omega C - \frac{1}{\omega L} \right).$$

The condition of minimum admittance with variation of frequency occurs when the susceptance $\omega C - \dfrac{1}{\omega L}$ vanishes, at the angular frequency ω_r given by

$$\omega_r^2 L C = 1.$$

This gives a voltage resonance. The complex admittance is then real and equal to G.

At frequencies below resonance the susceptance is negative, and above resonance it is positive. As the frequency falls below resonance the inductor becomes increasingly dominant, and eventually behaves as a dead short when dc conditions are approached. At frequencies above resonance

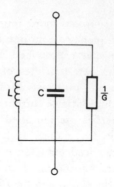

Fig. 8.7 Parallel circuit with conductance shunt

the capacitor is dominant, and provides increasingly large susceptance as the frequency tends to infinity.

The voltage v developed across the circuit is given by

$$|v|^2 = |i|^2 \Big/ \left[G^2 + \left(\omega C - \frac{1}{\omega L} \right)^2 \right],$$

where i is the current provided by the constant-current source. The variation of $|v|$ with frequency is shown in Fig. 8.8. At resonance $|v|$ reaches a maximum value equal to $\dfrac{|i|}{G}$. The magnitude of the current in the inductor is then

$$|i_L| = |i|/G|j\omega_r L|.$$

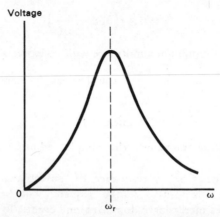

Fig. 8.8 Voltage-frequency characteristic of parallel circuit with conductance shunt

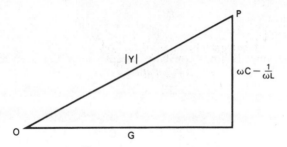

Fig. 8.9 Argand diagram for complex admittance of parallel circuit

The current gain at resonance is therefore

$$\left|\frac{i_L}{i}\right| = \frac{1}{G\omega_r L} = \frac{\omega_r C}{G}.$$

At resonance all the source current flows through the resistor. The currents in L and C are in antiphase with each other and of equal amplitude, and constitute a circulatory current flowing in these two components in virtual isolation from the supply circuit.

We shall now derive an expression for the difference in frequency between the half-power points. In the Argand diagram of Fig. 8.9 the point P represents the complex admittance Y. It will be recalled that the arguments of complex admittance and impedance are equal in magnitude and opposite in sign (section 6.10.2), and that the argument of Y is $-\phi$, if ϕ is the phase lead of voltage over current. The power factor of the circuit is

$$\cos \phi = \cos(-\phi) = \cos(\arg Y) = \frac{G}{|Y|}.$$

The electric power delivered into the circuit is

$$P = \frac{i_0^2}{2|Y|}\cos \phi = \frac{i_0^2 G}{2|Y|^2}.$$

At resonance $|Y| = G$, and the power attains a maximum value

$$P_{max} = \frac{i_0^2}{2G}.$$

At the half-power points

$$P = \frac{P_{max}}{2},$$

so that

$$\frac{i_0^2 G}{2|Y|^2} = \frac{i_0^2}{4G},$$

where the assumption of a constant-current source has been maintained. Thus

$$|Y| = \sqrt{2}G,$$

and

$$2G^2 = |Y|^2 = G^2 + \left(\omega C - \frac{1}{\omega L}\right)^2,$$

or

$$\left(\omega C - \frac{1}{\omega L}\right) = \pm G.$$

It is noticeable that this calculation has much in common with the similar analysis for the LCR series circuit. We see that at the half-power points the magnitudes of the susceptance and conductance are equal, so that the phase angle is $\pm(\pi/4)$ radians, and the power factor is $1/\sqrt{2}$.

It is quite easy to solve the last equation as a quadratic in ω, obtaining the roots

$$2\omega_1 C = -G + \sqrt{\left(G^2 + \frac{4C}{L}\right)},$$

and

$$2\omega_2 C = G + \sqrt{\left(G^2 + \frac{4C}{L}\right)}.$$

If the conductance is small, it is evident here that the mean value ω_m of the angular frequencies ω_1 and ω_2 is given approximately by

$$\omega_m^2 LC = 1.$$

The difference in frequencies between the roots is given by

$$2(\omega_2 - \omega_1)C = 2G,$$

so that for the quality factor

$$\frac{\omega_r}{\omega_2 - \omega_1} = \frac{\omega_r C}{G} = \frac{1}{G\omega_r L}.$$

It is noticeable that the quality factor and the current gain at resonance are equal.

The amplitude of the voltage at the half-power points is

$$v_0 = \frac{i_0}{|Y|} = \frac{i_0}{\sqrt{2}G},$$

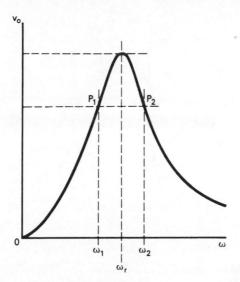

Fig. 8.10 Half-power points for parallel circuit

and this is $1/\sqrt{2}$ of the voltage developed across the circuit at resonance. The half-power points are indicated at P_1 and P_2 in Fig. 8.10.

8.5.2   *Parallel combination with series resistances*

In the circuit of Fig. 8.11, series resistance appears in both branches of the LC parallel combination. The complex admittance is

$$Y = \frac{1}{r_L + j\omega L} + \frac{1}{r_C + \dfrac{1}{j\omega C}}.$$

It is easy to guess that for a particular frequency the admittance must pass through a minimum value. It is difficult to derive an exact formula for this frequency, and it is therefore usual to assume that for this condition of voltage resonance the complex admittance is purely real. This is found to be an acceptable approximation if the series resistances are small. We shall therefore take as our criterion for resonance that the circuit susceptance will be zero, in which case

$$\frac{\omega_r L}{r_L^2 + \omega_r^2 L^2} - \frac{\dfrac{1}{\omega_r C}}{r_C^2 + \dfrac{1}{\omega_r^2 C^2}} = 0,$$

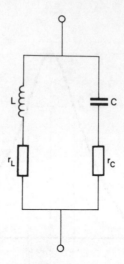

Fig. 8.11 Parallel circuit with series resistances

where ω_r is the angular frequency for resonance. This relation gives on rearrangement

$$\omega_r^2 LC = \frac{\dfrac{L}{C} - r_L^2}{\dfrac{L}{C} - r_C^2}.$$

If the resistances are small, then

$$\omega_r^2 LC = 1, \quad \text{approximately.}$$

The admittance at resonance is

$$Y = \frac{r_L}{r_L^2 + \omega_r^2 L^2} + \frac{r_C}{r_C^2 + \dfrac{1}{\omega_r^2 C^2}}.$$

Neglecting the squares of the resistances in each denominator,

$$Y = \frac{r_L}{\omega_r^2 L^2} + r_C \omega_r^2 C^2,$$

which in turn gives approximately

$$Y = \frac{r_L + r_C}{\omega_r^2 L^2}.$$

Putting
$$r = r_L + r_C,$$

which is the total series resistance in the parallel circuit, the impedance at resonance is

$$\frac{\omega_r^2 L^2}{r} = \frac{L}{Cr}.$$

If current i is drawn from the source, then the voltage which develops across the circuit at resonance is

$$\frac{Li}{Cr}.$$

The modulus of the current which flows in L is therefore approximately

$$|i_L| = \frac{Li}{Cr|j\omega_r L|},$$

so that the current gain at resonance is

$$\left|\frac{i_L}{i}\right| = \frac{1}{\omega_r Cr} = \frac{\omega_r L}{r}.$$

If the series resistances are small, the currents in L and C are large and almost equal in amplitude. Furthermore, the two are approximately in antiphase, and together constitute a large circulatory current. The system is sometimes compared to a rotating flywheel. The small energy loss in the series resistors is analogous to the loss by friction in the bearings. The current drawn from the supply maintains the mean energy of the system at a constant level.

The general variation with frequency of the voltage developed across the circuit does not greatly depend on whether shunt or series representation is used for circuit resistance, although in the latter case the voltage does not quite fall to zero as the extreme values of frequency are approached.

It can be shown by means of further rather tedious calculations that with the series representation of resistance used here, there is once more approximate equality of the current gain at resonance and the quality factor. Thus

$$\frac{\omega_r}{\omega_2 - \omega_1} = \frac{1}{\omega_r Cr} = \frac{\omega_r L}{r},$$

where ω_2 and ω_1 are the angular frequencies at the half-power points. For high selectivity it is therefore again necessary to minimise energy loss.

Short, thick, connecting wires should be provided for the two links between coil and capacitor. This is important partly to minimise stray reactance, and partly because ohmic loss produced by the heavy circulatory current may otherwise significantly reduce the quality factor. Relatively thin wire can be used for the connections between the tuned circuit and source.

Examples

1. A pure 60 mH inductor is joined in series with a pure capacitor C. A pure resistor R is connected in parallel with C. The arrangement behaves as a pure resistance at an angular frequency of 8×10^4 rad s^{-1}. In the absence of R the circuit is series resonant at an angular frequency of 10^5 rad s^{-1}. Determine the value of R.

2. A series LCR circuit consists of an inductive coil L having an effective series resistance of 20 Ω, a 200 pF pure capacitor, and a 980 Ω resistor. If the arrangement is series resonant at an angular frequency of 5×10^6 rad s^{-1}, determine the value of L and the fraction of the applied voltage which appears across the capacitor at resonance.

3. In a series LCR circuit $L = 50$ mH and $R = 10$ Ω. If the 1 mV source is of low impedance, determine the approximate angular frequency separation of the half-power points, and the power consumed at these points.

4. A good quality capacitor C is connected in series with a 1 mH inductive coil which has series resistance r. The arrangement is series resonant at an angular frequency of 10^6 rad s^{-1}, and gives a quality factor of 500. Obtain values for C and r.

5. The low-impedance source of a Q-meter provides an emf of 100 mV. A high-impedance voltmeter is connected across the tuning capacitor. In the absence of any external admittance the voltmeter displays a maximum reading of 10 V when the capacitor is adjusted to a value of 1 nF, the angular frequency of the source being 5×10^6 rad s^{-1}. Obtain values for L and the effective series resistance of the circuit.

6. In the absence of any external admittance, the Q exhibited at resonance by a Q-meter circuit is 500 at an angular frequency of 5×10^6 rad s^{-1}, the value of the tuning capacitor C being 400 pF.

When a pure resistor R is connected across C and the latter is retuned, the Q is found to have fallen to 400. Obtain values for the series inductance and resistance of the Q-meter circuit, and the approximate resistance of R.

7.   A pure inductor L is shunted by a pure capacitor C, and these two components are in turn shunted by a pure resistor R. If $C = 500$ pF and $R = 100$ kΩ, give the value of L for parallel resonance at an angular frequency of 10^6 rad s^{-1}, and the voltage which then develops across the circuit if a current of 1 μA is drawn from the source. Determine the magnitude of the current which circulates in the reactances.

8.   In a parallel-tuned circuit both reactive components L and C have series resistance. Obtain an approximate value for the resonant angular frequency if $L = 100$ μH, $C = 400$ pF, and the total series resistance in the closed loop formed by the reactors is 10 Ω. If an ac current of 10 μA is drawn from the source, what approximate current circulates in the loop? What are the quality factor and circuit impedance at resonance?

9.   In a parallel-tuned circuit both reactive components have series resistance. The circuit behaves as a 10 kΩ resistance at the resonant frequency of 1 MHz, the Q being 10 π. Obtain approximate values for the inductance and capacitance, and the total series resistance in the loop. If a supply voltage of 5 mV is provided at resonance, what approximate total current is drawn and what current circulates in the loop?

Transformers

9.1 General considerations

Basic aspects of transformer behaviour were discussed in an earlier chapter (5.3), and equations were established relating time-dependent primary and secondary currents. We shall shortly be re-examining these in the context of ac conditions.

A common transformer circuit is shown in Fig. 9.1(*a*). The dc isolation of the primary and secondary coils is a useful incidental advantage. An autotransformer is illustrated in Fig. 9.1(*b*). Such an arrangement obviously cannot provide electrical isolation, but it is otherwise as effective as the previous type.

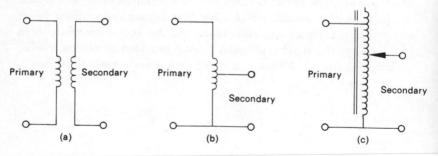

Fig. 9.1 Transformer circuits

If an iron core is provided, the self and mutual inductances of the coils are sufficiently high to permit use at frequencies as low as that of the ac mains. The transformer of Fig. 9.1(*c*) is designed for direct connection across the ac mains. The output voltage is continuously variable, and a step-up ratio in excess of unity is available. Where isolation from the mains supply is desired for reasons of safety or convenience, an ordinary isolating transformer must be interposed between the autotransformer and the supply.

9.1.1 Voltage ratio

Suppose that the secondary of an air-cored transformer is open-circuited, and that in both windings the effects of ohmic resistance and capacitance between turns may be neglected. Let an ac voltage be applied to the terminals of the primary coil. No current flows in the secondary, so that conditions in the primary are undisturbed by the presence of the secondary. The emf induced in the primary by the primary current therefore just equals the applied voltage.

The close mutual proximity of the turns of a conventional transformer ensures that roughly the same emf is induced in each. The emf available at the terminals of the secondary coil and the voltage applied to the primary are therefore approximately in the ratio of the numbers of turns on the secondary and primary coils respectively. Thus

$$\frac{\text{output voltage}}{\text{input voltage}} = \frac{\text{number of secondary turns}}{\text{number of primary turns}}.$$

For many transformers this relation is no more than a useful guide, especially at high frequencies. It becomes very much more accurate for iron-cored transformers and at low frequencies, where the principal source of error is internal voltage drop in the ohmic resistance of the windings. Evidently the number of secondary turns exceeds the number of primary turns for a step-up transformer, and in general a very wide range of voltage ratios can be produced by suitable design.

The efficiency of a transformer is the ratio of power output to power input, and at mains frequencies it can exceed 95%. For the ideal of perfect power transference, and for power factors in the vicinity of unity, the product of voltage and current will be the same for both windings. If the voltage is stepped up, the current is stepped down in the same ratio, so that an impedance transformation is effected in proportion to the squares of the numbers of turns. Transformers are often used primarily with a view to effecting an impedance match.

9.1.2 Coupling factor

The coupling factor k of a transformer is defined by the relation

$$M^2 = kL_1L_2,$$

where L_1 and L_2 are the self inductances of the primary and secondary coils respectively. In an alternative definition k^2 appears in place of k. The coupling factor is a measure of the extent of the interaction between the primary and secondary coils. The theoretical maximum value is unity, and it approaches zero when the separation between the coils is large. For iron-cored low-frequency transformers the coupling factor is close to unity, but it is generally much lower than this for transformers designed for high-frequency operation. It is somewhat easier to attain high values of coupling factor for autotransformers than for electrically separated coils.

9.1.3   Loss mechanisms

Inefficient transference of power from primary to secondary is due principally to magnetic losses in the material of the core, and to ohmic resistance in the copper windings. Ohmic losses rise with frequency, as skin effect causes the current to be concentrated increasingly near the surface of the wire.

At high frequencies an additional effect is power loss by radiation, in the manner of an aerial.

For low-frequency operation a soft-iron core is essential to provide adequate coupling. Eddy currents associated with emfs induced in the core are suppressed by constructing it with thin oxidised laminations arranged with planes perpendicular to the direction of the induced current flow, so that the electrical resistance in the current path is very high. Such cores are employed at frequencies up to at least 20 kHz. Above 100 kHz dust (bonded) cores or air cores are employed.

The use of ferrous cores gives rise to nonlinearities, and these add considerably to the complexities of theoretical analysis. An obvious symptom is the distortion of sinusoidal current and voltage waveforms caused by the dependence of the self and mutual inductances on the instantaneous currents in the coils. A full treatment would involve an unprofitable excursion into the study of magnetic properties of materials, and would need to take account of additional effects such as stray capacitance between turns. The relatively simple analyses presented in this chapter are valid only for an idealised air-cored transformer, but should give helpful indications of the behaviour and potentialities of real transformers.

9.1.4   Open-circuit secondary voltage

Suppose an alternating current i_1 flows in the primary of the transformer illustrated in Fig. 9.2. If the secondary circuit is open, the emf v_2 induced in the secondary coil appears unmodified at the output terminals. The relation between the primary current and secondary emf is then

$$v_2 = M \frac{di_1}{dt} = j\omega M i_1,$$

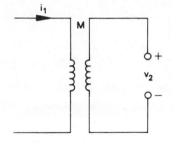

Fig. 9.2 Transformer with open-circuited secondary

where time-dependence of the form $\exp j\omega t$ has been assumed for i_1. Evidently, if M is positive the secondary voltage leads the primary current by $\pi/2$ radians.

9.1.5 Basic equations

Let a low-impedance source of emf v_1 be connected to the primary of a transformer, as shown in Fig. 9.3. A load which behaves as a complex impedance Z at the source frequency is connected to the terminals of the secondary. The instantaneous polarity of the source and the directions shown for the currents conform to the sign convention detailed in section 5.3.9.

Applying Kirchhoff's second law to the primary and secondary circuits in turn,

$$v_1 = R_1 i_1 + L_1 \frac{di_1}{dt} + M \frac{di_2}{dt},$$

and

$$0 = R_2 i_2 + L_2 \frac{di_2}{dt} + M \frac{di_1}{dt} + v_2.$$

The operator $\dfrac{d}{dt}$ can be replaced by the factor $j\omega$ throughout, so that these relations can be written alternatively:

$$v_1 = R_1 i_1 + j\omega L_1 i_1 + j\omega M i_2,$$

and

$$0 = R_2 i_2 + j\omega L_2 i_2 + j\omega M i_1 + Z i_2.$$

Let the total series complex impedances of the primary and secondary circuits be Z_1 and Z_2 respectively. Then

$$Z_1 = R_1 + j\omega L_1,$$

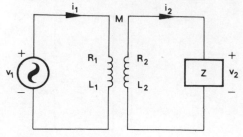

Fig. 9.3 Transformer with complex load Z

and

$$Z_2 = R_2 + j\omega L_2 + Z.$$

The two circuit relations can now be written as

$$v_1 = Z_1 i_1 + j\omega M i_2,$$

and

$$0 = Z_2 i_2 + j\omega M i_1.$$

These can easily be arranged in the form

$$\frac{i_1}{Z_2} = \frac{i_2}{-j\omega M} = \frac{v_1}{Z_1 Z_2 + \omega^2 M^2}.$$

9.1.6 *Input impedance*

The impedance presented by the primary to the source is

$$\frac{v_1}{i_1} = Z_1 + \frac{\omega^2 M^2}{Z_2}.$$

The first term on the right-hand side of this equation is the impedance of the primary coil. The second term results from interaction with the secondary, and is sometimes called the *reflected impedance*. If the secondary were open-circuited, the input impedance would be contributed only by the primary, and would be inductive in character.

Now suppose $|Z|$ is finite. Multiply numerator and denominator of the last term in the above equation by Z_2^*, the complex conjugate of Z_2. Then

$$\frac{v_1}{i_1} = Z_1 + \frac{\omega^2 M^2}{Z_2 Z_2^*} Z_2^*.$$

The quantity $Z_2 Z_2^*$ is real. It is apparent that the reflected impedance contributes additional input series resistance

$$\frac{\omega^2 M^2}{Z_2 Z_2^*} \, \text{Re} \, (Z_2^*),$$

and additional input series reactance

$$\frac{\omega^2 M^2}{Z_2 Z_2^*} \, \text{Im} \, (Z_2^*).$$

The first of these quantities is essentially positive. Now unless Z is capacitive, Z_2 will be inductive in character. It follows that $\text{Im} \, (Z_2^*) < 0$, and coupling with the secondary reduces the input reactance. The overall effect is therefore to bring the primary current more closely into phase with the supply voltage. In general, also, the actual magnitude of the primary current is increased, because the diminution in reactance usually outweighs the increase in resistance. These two effects act in combination to produce an increase in power consumption.

For mains transformers, with the secondary open-circuited the primary reactance is usually considerably greater than the primary resistance, and only a small and nearly wattless current flows. This is sometimes called the magnetising current.

9.1.7   Resistive loading

If the secondary load Z is purely real and equal to R, then

$$Z_2 = R_2 + j\omega L_2 + R.$$

Put

$$R_2' = R_2 + R.$$

The effective input impedance is

$$\frac{v_1}{i_1} = R_1 + j\omega L_1 + \frac{\omega^2 M^2}{Z_2 Z_2^*}(R_2' - j\omega L_2).$$

As an exercise we shall investigate whether this can be purely real. We shall require

$$\omega L_1 = \frac{\omega^2 M^2}{Z_2 Z_2^*} \, \omega L_2.$$

It follows that

$$Z_2 Z_2^* = \omega^2 M^2 \frac{L_2}{L_1},$$

or

$$R_2'^2 + \omega^2 L_2^2 = k\omega^2 L_2^2,$$

where k is the coupling factor. Thus

$$R_2'^2 = -\omega^2 L_2^2 (1 - k).$$

The right-hand side of this relation is essentially negative. Therefore when the secondary load is purely resistive, the effective input impedance can never be real, and the input reactance will always be positive.

9.1.8 Equivalent primary circuit

With the aid of the relation

$$M^2 = kL_1 L_2,$$

the effective primary input impedance can be written

$$\frac{v_1}{i_1} = Z_1 + \frac{\omega^2 k L_1 L_2}{Z_2}.$$

If we put

$$Z_1 = R_1 + j\omega L_1,$$

and

$$Z_2 = R_2 + j\omega L_2 + Z,$$

it is then quite easy to obtain the alternative form

$$\frac{v_1}{i_1} = R_1 + j\omega L_1(1 - k) + 1 \Big/ \left[\frac{1}{j\omega L_1 k} + \frac{L_2}{L_1 k (R_2 + Z)} \right].$$

This can be represented by the equivalent circuit of Fig. 9.4. The quantity $L_1(1 - k)$ is called the *leakage inductance* of the primary. It may be thought of as that part of the primary inductance which is not involved in the creation of secondary emf. Using this type of equivalent circuit it is

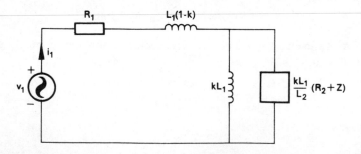

Fig. 9.4 Equivalent primary circuit for air transformer

easy to allow for additional effects, such as shunt capacitance between turns and core losses, by adding appropriate impedances.

9.1.9   Worked example

We shall suppose that a purely capacitive load C is attached to the secondary coil of a transformer, and that the output voltage is required to be in antiphase with the supply voltage.

The output voltage is

$$v_2 = i_2 Z = \frac{i_2}{j\omega C}.$$

It follows from the relations derived in section 9.1.5 that

$$\frac{v_1}{v_2} = \frac{j\omega C v_1}{i_2} = -(Z_1 Z_2 + \omega^2 M^2)\,\frac{C}{M}.$$

Evidently v_1 and v_2 will be precisely in antiphase if the quantity $Z_1 Z_2$ is wholly real. Then

$$\text{Im}\,(Z_1 Z_2) = \text{Im}\left[\left(R_1 + j\omega L_1\right)(R_2 + j\omega L_2 + \frac{1}{j\omega C})\right] = 0.$$

Thus

$$\omega^2 C(R_1 L_2 + R_2 L_1) = R_1.$$

For a symmetrical transformer $R_1 = R_2$, and $L_1 = L_2 = L$, say, so that the required condition is then simply

$$2\omega^2 LC = 1.$$

9.2   Tuned transformer

Transformer coils are essentially inductive in character, and the provision of suitably arranged capacitive tuning can therefore confer the advantages of selectivity and maximum power transference. The transformer illustrated in Fig. 9.5 is provided with tuning capacitors connected to the primary and secondary coils. If a constant-current ac source is connected in parallel with C_1, it is presented with a suitably high input impedance at resonance. A constant-voltage source, on the other hand, would be connected in series with the closed primary circuit, so that a low input impedance is provided at resonance. Both situations are conducive to heavy current flow in the primary coil, and are therefore favourable to the development of voltage gain.

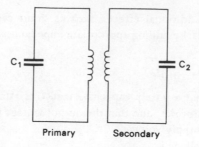

Primary   Secondary

Fig. 9.5 Transformer provided with tuning capacitors

A similar choice is available for the siting of the load. A high impedance load should be connected in parallel with C_2 to exploit the voltage amplification of the secondary at resonance. A low impedance load should be connected in series to exploit the current gain at resonance.

As with coupled tuned systems generally, the behaviour is more complicated than consideration of either system alone might suggest. When the coupling factor is low it is found that with both circuits tuned to the signal frequency a maximum of output voltage occurs at that frequency. But if the coupling is increased beyond a certain critical value, maxima in output develop at two frequencies displaced to opposite sides of the original maximum.

A complete analysis of the behaviour of a transformer with tuned primary and secondary is very involved. Fortunately the principal features can be exposed without a great deal of difficulty. We shall look first at the effects of adjustment of the tuning of the primary, and then of the secondary. Attention will be restricted to circuits suited for constant-voltage sources and high impedance loads.

For alternative treatments of tuning the reader is referred to the book list given at the end of the chapter.

9.2.1   *Primary tuning*

The air transformer of Fig. 9.6 is provided with series capacitance tuning in the primary circuit. The voltage gain is $\left|\dfrac{v_2}{v_1}\right|$. Using the relations derived in section 9.1.5, we have the following expression for the reciprocal of this quantity:

$$\left|\frac{v_1}{v_2}\right| = \left|\frac{j\omega C_2 v_1}{i_2}\right| = \left|\frac{j\omega C_2}{j\omega M}(Z_1 Z_2 + \omega^2 M^2)\right|.$$

Here Z_1 and Z_2 are the series complex impedances of the primary and secondary circuits respectively. We shall represent these as

$$Z_1 = R_1 + jX_1,$$

and

$$Z_2 = R_2 + jX_2,$$

where X_1 and X_2 are real.

If C_1 is adjusted for maximum voltage gain, the condition which results satisfies

$$\frac{\partial}{\partial X_1} \left| \frac{v_1}{v_2} \right| = 0.$$

It will however be easier to examine the consequences of the equivalent relation

$$\frac{\partial}{\partial X_1} \left| \frac{v_1}{v_2} \right|^2 = 0.$$

Now

$$\left| \frac{v_1}{v_2} \right|^2 = \left(\frac{C_2}{M} \right)^2 [(R_1R_2 - X_1X_2 + \omega^2M^2)^2 + (R_1X_2 + R_2X_1)^2].$$

Thus we require

$$2(R_1R_2 - X_1X_2 + \omega^2M^2)(-X_2) + 2(R_1X_2 + R_2X_1)R_2 = 0.$$

This reduces easily to

$$\omega^2M^2X_2 = X_1(R_2^2 + X_2^2) = X_1|Z_2|^2,$$

or

$$\left(\frac{\omega M}{|Z_2|} \right)^2 = \frac{X_1}{X_2}.$$

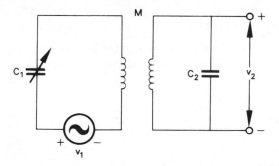

Fig. 9.6 Air transformer provided with capacitive primary tuning and capacitive load

This formula is unchanged if an unspecified impedance occupies the position of the capacitor C_2.

9.2.2   Input impedance of tuned primary

The impedance presented by the primary circuit to the source is

$$\frac{v_1}{i_1} = Z_1 + \frac{\omega^2 M^2}{Z_2} = Z_1 + \frac{\omega^2 M^2}{|Z_2|^2} Z_2^*,$$

where Z_2^* is the complex conjugate of Z_2.

Let the primary circuit be tuned to produce maximum voltage gain. Then with the aid of the relation derived in the previous section we have

$$\frac{v_1}{i_1} = Z_1 + \frac{X_1}{X_2} Z_2^* = (R_1 + jX_1) + \frac{X_1}{X_2} (R_2 - jX_2).$$

Thus

$$\frac{v_1}{i_1} = R_1 + \frac{X_1}{X_2} R_2.$$

The input impedance of the tuned primary will therefore be real, and in general low. With low input impedance a heavy primary current flows, and this encourages the development of a large output voltage.

The voltage gain which is obtained in this way can be quite large. If the primary and secondary circuit component values have similar magnitudes, then the primary and secondary currents will also have similar magnitudes. The voltage gain is a consequence of the step-up in impedance associated with the transition from the series-fed primary to the parallel-extracted secondary.

For a symmetrical transformer arrangement $X_1 = X_2$, and the input impedance is exactly $R_1 + R_2$.

9.2.3   Tuning of secondary

Now let the secondary circuit be tuned for maximum voltage gain by adjustment of the capacitor C_2. Let the series inductance of the secondary be L_2 (Fig. 9.7). Then

$$X_2 = \omega L_2 - \frac{1}{\omega C_2},$$

and

$$\frac{\partial X_2}{\partial C_2} = \frac{1}{\omega C_2^2}.$$

We shall investigate the condition

$$\frac{\partial}{\partial X_2} \left| \frac{v_1}{v_2} \right|^2 = 0,$$

where, as before (section 9.2.1):

$$\left| \frac{v_1}{v_2} \right|^2 = \left(\frac{C_2}{M} \right)^2 [(R_1 R_2 - X_1 X_2 + \omega^2 M^2)^2 + (R_1 X_2 + R_2 X_1)^2].$$

Therefore

$$M^2 \frac{\partial}{\partial X_2} \left| \frac{v_1}{v_2} \right|^2 = 2C_2 \frac{\partial C_2}{\partial X_2} [(R_1 R_2 - X_1 X_2 + \omega^2 M^2)^2 + (R_1 X_2 + R_2 X_1)^2]$$

$$+ 2C_2^2 [(R_1 R_2 - X_1 X_2 + \omega^2 M^2)(-X_1) + (R_1 X_2 + R_2 X_1)(R_1)].$$

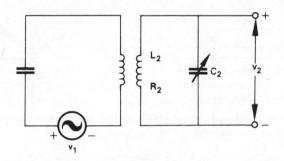

Fig. 9.7 Air transformer with capacitive secondary tuning

It follows that

$$\omega C_2 [(R_1 R_2 - X_1 X_2 + \omega^2 M^2)^2 + (R_1 X_2 + R_2 X_1)^2] + X_2 |Z_1|^2 - \omega^2 M^2 X_1 = 0.$$

If resistance terms are neglected, this reduces to

$$\omega C_2 (\omega^2 M^2 - X_1 X_2)^2 = X_1 (\omega^2 M^2 - X_1 X_2).$$

One root of this equation is

$$\omega^2 M^2 - X_1 X_2 = \frac{X_1}{\omega C_2}.$$

This root is not of significance here, for it is easily reduced to the form

$$\frac{\omega M^2}{L_2} = X_1,$$

a condition which is unaffected by the value of C_2.

A second root is

$$\omega^2 M^2 = X_1 X_2.$$

It can be seen that in the absence of loss this is identical with the condition obtained in section 9.2.1 for tuning of the primary.

9.2.4   Symmetrical tuned transformer

The conditions for maximum voltage gain with variation of tuning capacitance have been found to be closely similar for the primary and secondary circuits. This suggests that the primary and secondary coils should be made identical, and that equal tuning capacitances should be provided. We shall now investigate the variation with frequency of the behaviour of the resulting symmetrical arrangement.

In section 9.2.1 we used the relation

$$\left|\frac{v_1}{v_2}\right|^2 = \left(\frac{C_2}{M}\right)^2 [(R_1 R_2 - X_1 X_2 + \omega^2 M^2)^2 + (R_1 X_2 + R_2 X_1)^2]$$

as a measure of the voltage gain. This could be differentiated with respect to ω with a view to discovering the frequency or frequencies at which maxima occur. Unfortunately the equation so obtained is quite complicated, and it is found to be very much easier to study instead the behaviour of the secondary current i_2. Now

$$i_2 = j\omega C_2 v_2,$$

and over small frequency ranges the factor ωC_2 does not vary very considerably. Thus useful indications can be derived in this way as to the behaviour of the secondary voltage v_2. With a little rearrangement of our earlier equation we have

$$\omega^2 M^2 \left|\frac{v_1}{i_2}\right|^2 = (R_1 R_2 - X_1 X_2 + \omega^2 M^2)^2 + (R_1 X_2 + R_2 X_1)^2.$$

For the symmetrical transformer this becomes

$$\omega^2 M^2 \left|\frac{v_1}{i_2}\right|^2 = (R^2 - X^2 + \omega^2 M^2)^2 + 4R^2 X^2$$

$$= (R^2 + X^2)^2 + 2\omega^2 M^2 (R^2 - X^2) + \omega^4 M^4.$$

Further,

$$X = \omega L - \frac{1}{\omega C},$$

so that

$$\omega \frac{\partial X}{\partial \omega} = \omega L + \frac{1}{\omega C}.$$

Now

$$2\omega M^2 \left|\frac{v_1}{i_2}\right|^2 + \omega^2 M^2 \frac{\partial}{\partial \omega} \left|\frac{v_1}{i_2}\right|^2 = 4(R^2 + X^2)X \frac{\partial X}{\partial \omega} + 4\omega M^2 (R^2 - X^2)$$

$$-4\omega^2 M^2 X \frac{\partial X}{\partial \omega} + 4\omega^3 M^4.$$

Putting

$$\frac{\partial}{\partial \omega} \left|\frac{v_1}{i_2}\right|^2 = 0,$$

this reduces quite easily to the form

$$\left(R^2 + X^2 - \omega^2 M^2\right)\left(R^2 + X^2 + \omega^2 M^2 - 2\omega^2 L^2 + \frac{2}{\omega^2 C^2}\right) = 0.$$

Evidently there are two solutions. One of these is

$$R^2 + X^2 = \omega^2 M^2.$$

This relation is a condition for maximum gain with variation of frequency for a symmetrical transformer arrangement. It is seen to be indistinguishable from the conditions for maximum voltage gain obtained in sections 9.2.1 and 9.2.3 for separate capacitance tuning of the primary and secondary respectively, provided circuit resistance is neglected in the latter situation. The other solution is

$$R^2 + X^2 + \omega^2 M^2 - 2\omega^2 L^2 + \frac{2}{\omega^2 C^2} = 0.$$

We shall examine these solutions separately.

9.2.5 The principal solution

The second of the above solutions can be written in the alternative form

$$R^2 - \frac{2L}{C} + \omega^2 M^2 - \omega^2 L^2 + \frac{3}{\omega^2 C^2} = 0,$$

or

$$\omega^4 L^2 (1 - k) + \frac{2\omega^2 L}{C}\left(1 - \frac{CR^2}{2L}\right) - \frac{3}{C^2} = 0.$$

The solution of this quadratic in ω^2 is

$$\omega^2 LC(1-k) = -\left(1 - \frac{CR^2}{2L}\right) \pm \sqrt{\left[\left(1 - \frac{CR^2}{2L}\right)^2 + 3(1-k)\right]}.$$

The quantity under the square root sign is obviously always real if resistive effects are small. Since k cannot exceed unity the left-hand side of the equation is positive. Thus only the positive choice of sign on the right-hand side is acceptable. If further we use the substitution

$$\left(1 - \frac{CR^2}{2L}\right) = n\sqrt{(1-k)},$$

the solution becomes

$$\omega^2 LC\sqrt{(1-k)} = \sqrt{(n^2+3)} - n.$$

Evidently this solution exists for all n and k. But we shall shortly discover that the magnitudes of n and k determine whether the solution represents a maximum or minimum for the secondary current i_2.

9.2.6   The large coupling solutions

The other relation obtained in section 9.2.4 was

$$R^2 + X^2 = \omega^2 M^2,$$

or

$$R^2 + \left(\omega L - \frac{1}{\omega C}\right)^2 = \omega^2 M^2.$$

This is a quadratic in ω^2, and if we again use the substitution

$$\left(1 - \frac{CR^2}{2L}\right) = n\sqrt{(1-k)},$$

we obtain

$$\omega^2 LC\sqrt{(1-k)} = n \pm \sqrt{(n^2-1)}.$$

Both these roots will exist provided

$$n^2 \geqslant 1,$$

or

$$\left(1 - \frac{CR^2}{2L}\right)^2 \geqslant 1 - k.$$

Usually R^2 is small in comparison with the quantity $\dfrac{2L}{C}$. Thus an acceptable modification of the above inequality is

$$1 - \frac{CR^2}{L} \geqslant 1 - k, \quad \text{approximately,}$$

or

$$k \geqslant \frac{CR^2}{L}.$$

We can therefore identify a critical coupling factor

$$k_c = \frac{CR^2}{L}.$$

For smaller values a stationary secondary current occurs at only one frequency. For higher values there are three.

9.2.7 Response curves

Response curves for a symmetrical tuned transformer are reproduced in Fig. 9.8.

For ω tending to zero or infinity the secondary current falls to zero, since for each of these extremes the series impedance of the primary circuit is infinite. For low values of coupling factor there is a single maximum at a frequency given by the relation derived in section 9.2.5, and the height of this maximum rises as the coupling factor is increased. As the coupling factor passes through the critical value two new stationary values appear. These are maxima, which occur at the frequencies given by the relations derived in section 9.2.6. The stationary value which occurs at the condition of critical coupling factor, and out of which these develop, is now depressed to form a minimum.

We had for the symmetrical tuned transformer the relation

$$\omega^2 M^2 \left| \frac{v_1}{i_2} \right|^2 = (R^2 - X^2 + \omega^2 M^2)^2 + 4R^2 X^2.$$

At the outer maxima the roots for the angular frequency were given in section 9.2.4 by

$$R^2 + X^2 = \omega^2 M^2,$$

and it is a matter of simple algebra to show that if this condition is satisfied, then

$$\left| \frac{v_1}{i_2} \right| = 2R.$$

The outer maxima are therefore of equal height.

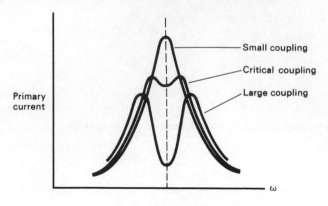

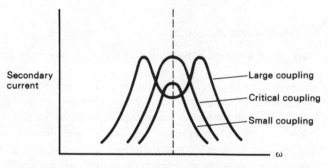

Fig. 9.8 Response curves for symmetrical tuned transformer (Redrawn from Duffin, W. J. (1965), *Electricity and Magnetism*, McGraw-Hill Book Company.)

The input impedance is

$$\frac{v_1}{i_1} = \frac{Z_1 Z_2 + \omega^2 M^2}{Z_2} = Z_1 + \frac{\omega^2 M^2}{Z_2 Z_2^*} Z_2^*$$

$$= Z_1 + \frac{\omega^2 M^2}{R^2 + X^2} Z_2^*.$$

It is apparent that at the outer maxima this reduces to

$$\frac{v_1}{i_1} = Z_1 + Z_2^*.$$

As Z_1 and Z_2 are identical, the input impedance is real and equal to $2R$. The results of the last two paragraphs indicate that the primary and secondary currents are equal at the outer maxima.

9.2.8   *Practical considerations*

If the coupling factor slightly exceeds the critical value, the response curve for the secondary current exhibits two closely spaced humps of equal height, with a shallow depression between. The transformer will in consequence provide a fairly level and quite narrow pass band, with a rapid fall-off on each side. In this situation it is especially useful in intermediate-frequency amplifiers, where a uniform response is required over a frequency band which may be at most a few per cent of the centre frequency, and where strong discrimination against signals of frequencies outside this band is essential.

Additional reading

DUFFIN, W. J., *Electricity and Magnetism*. McGraw-Hill, 1965.
FICH, S. and POTTER, J. L., *Theory of AC Circuits*. Prentice-Hall, 1958.
HARNWELL, G. P., *Principles of Electricity and Electromagnetism*. McGraw-Hill, 1949.
PAGE, L. and ADAMS, N. I., *Principles of Electricity*. Van Nostrand, 1969.

Examples

1.   The open-circuited secondary emf in an air transformer is 50 mV when the primary current is 10 mA at an angular frequency of 1000 rad s^{-1}. What is the mutual inductance of the transformer?

2.   The self inductance of each coil of a symmetrical air-cored 5 mH mutual inductor is 10 mH. The primary and secondary are connected in series, and an ac voltage of 30 mV is applied across the unconnected terminals. Give two possible values for the current which flows, the angular frequency being 10^4 rad s^{-1}. Ignore resistance effects.

3.   The primary of a transformer is connected to an ac source, the secondary being initially open-circuited. When the secondary circuit is closed through a load, it is found that the increase in the input resistance of the primary is equal to the fall in the input reactance. What is the argument of the complex impedance of the closed secondary circuit?

4.   The resistance of the secondary of a 50 mH air-cored mutual

inductor is 50 Ω. The secondary circuit is closed by a capacitor which would series-resonate with the self inductance of the secondary coil in the absence of coupling between primary and secondary. Determine the contribution made by the secondary circuit to the input impedance of the primary, the angular frequency being 1000 rad s^{-1}.

5. A pure capacitor C is connected as the secondary load of a symmetrical air-cored mutual inductor. If the self inductance of each winding is L, show that if

$$2\omega^2 LC = 1,$$

where ω is the angular frequency, then the supply and output voltages are precisely in antiphase.

   Find the value of C if the source frequency is 5 kHz, and $L = 5$ mH.

6. The self inductance and series resistance of both windings of a symmetrical air-cored transformer are 1 mH and 60 Ω respectively. The coupling factor is unity. A low-impedance 6 mV source of angular frequency 12×10^6 rad s^{-1} is connected to the primary. If the secondary is open-circuited, give approximate values for the primary current and the energy dissipation in the primary.

   If the secondary circuit is closed by a purely resistive load of 5940 Ω, determine the effective primary resistance and inductance.

10

Alternating current bridges

10.1   The bridge principle

It is likely the term 'bridge' referred originally to the bridging by a detector of the gap between two terminals in a circuit. It has since come to mean the measuring circuit as a whole.

The Wheatstone bridge used a battery as the source of steady current, and a dc current-measuring instrument as the detector. In the course of time the increasing need for the measurement of capacitance led to the development of ballistic bridges by Maxwell and others. The transient response was initiated by switching a battery in to or out of the circuit. The majority of ballistic bridges employed the basic Wheatstone configuration, which we shall call here Wheatstone-type. The detector was a ballistic galvanometer, which at balance remained undeflected. The absence of response normally indicated no current flowed at any time in the detector circuit, a condition described as *absolute balance.* A less demanding requirement was *aggregate balance,* where current might flow first one way through the galvanometer, and then the other, the net amount of charge passing through the instrument being zero.

The use of ac bridge sources dates from the work of Oberbeck in 1891. Detectors used in early work included vibration galvanometers, and later headphones. In the balanced bridge, the potential difference between the two detector terminals is zero at every instant in time. There is therefore equality of both amplitude and phase for the potential differences between these terminals and either terminal of the source. In general there will in consequence be two conditions of balance, and in a well-designed bridge independent adjustments can be made to satisfy each of these.

Despite their interchangeability, in Wheatstone-type bridges the source and detector do not possess a common terminal. If, for example, both the source and detector are designed to be operated with one terminal earthed, then one or other must be isolated electrically by means of a transformer. A more insoluble problem is the impossibility of permanently earthing on one side more than two of the four arms of the bridge, so that there are difficulties in screening. In spite of the inherent high sensitivity, this type of bridge has in consequence given way in the last three decades almost entirely to alternative basic designs.

At the present day the term bridge is used for any network using the null response principle, and in which the balance condition is undisturbed by interchange of source and detector.

10.2   Wheatstone-type bridges

The circuit of Fig. 10.1 is a general Wheatstone-type network. The complex impedances of the four bridge arms and the complex currents flowing in them are represented in the diagram by appropriate symbols. The source provides a pure sinusoidal voltage, and we shall suppose that the bridge component values can be adjusted so that no current flows in the detector. The potential difference between the ends of the detector branch will then be zero at every instant in time.

10.2.1   Balance conditions

No current flows in the detector branch at balance, so that

$$i_1 = i_2,$$

and

$$i_4 = i_3.$$

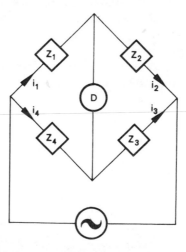

Fig. 10.1 General Wheatstone-type bridge network

In addition, the potential differences developed across Z_1 and Z_4 will be identical in amplitude and phase, as will those across Z_2 and Z_3. It follows therefore that

$$Z_1 i_1 = Z_4 i_4,$$

and

$$Z_2 i_2 = Z_3 i_3.$$

The above relations give

$$\frac{Z_1}{Z_4} = \frac{Z_2}{Z_3}.$$

This result is seen to be consistent with the familiar balance condition of the purely resistive Wheatstone bridge. It is unaffected by the internal impedance of the source, or by the magnitude of the voltage which that provides. The equation can be rearranged in several ways, and with a given ac bridge it is always worth a little care in selecting the most convenient form.

Because complex quantities are involved, two balance conditions will in general emerge. It is usual to express complex impedance in terms of resistance and reactance, which suggests that the real and imaginary parts of the equation should be separately equated. An equally valid procedure involves comparison of the moduli and arguments of the two sides of the equation. The balance conditions obtained in the two ways differ in form but are necessarily analytically equivalent.

When adjustments are made in the values of bridge components with a view to satisfying the two conditions of balance, the variables chosen should preferably be quite independent in their effect. In low-frequency bridges it is cheaper and easier to alter resistance values, but at high frequencies variation of reactance is considered to give greater accuracy.

10.2.2   Owen bridge

This bridge was devised in 1915 as a means of measuring self inductance. In the circuit of Fig. 10.2 L is the unknown inductance. Part of the resistance r is contributed by the inductor.

The general balance condition of 10.2.1 gives for this bridge

$$\frac{r + j\omega L}{R_0} = \left(R + \frac{1}{j\omega C} \right) j\omega C_0.$$

Equating reals,

$$\frac{r}{R_0} = \frac{C_0}{C},$$

and equating imaginaries:

$$\frac{L}{C_0} = RR_0.$$

These are the two conditions of balance.

The bridge is intended for low-frequency measurements, and there is therefore no advantage in providing variable capacitance in any bridge arm. Resistors r and R are obvious choices for variables, enabling the two conditions of balance to be satisfied independently. The first condition gives a value for the ohmic resistance of the coil, for which a fairly low standard of accuracy is usually acceptable. Allowance must be made for resistance contributed by the variable resistor in the same arm. The inductance is measured in terms of the capacitor C_0, which must therefore be a good quality capacitor of accurately known capacitance. For the same reason the components R and R_0 are good quality decade resistance boxes. The sensitivity of the measurement is diminished by the presence of the variable resistor in the same arm as the unknown, and r should therefore be kept small, consistent with a sufficiently large value of C being available to achieve balance.

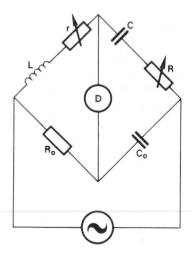

Fig. 10.2 Circuit of Owen bridge

It is a good principle in bridge measurements to decide at the outset which components in the bridge are of most significance. In the bridge under consideration r is kept small, and C is in consequence large. Neither affects the sensitivity to any marked extent, and neither appears in the important second condition of balance. The components of significance are L, R, C_0 and R_0. All are involved in the second balance condition. We shall call these the dominant components. It is noticeable that each bridge arm contains one.

The sensitivity may be found to be insufficient during preliminary adjustments. We are concerned here with the measurement of L in terms of the variable resistance R, and the relevant aspect of the sensitivity is the magnitude of the response of the detector to a given small fractional change in R when the bridge is in the vicinity of balance. Obviously there is benefit in using as sensitive a detector as possible, and as large a generator voltage as can be tolerated by the bridge components. As with the resistive Wheatstone network, the sensitivity is further enhanced if the impedances of the dominant components are adjusted to be each of about the same order of magnitude. This procedure can usually be applied with considerable benefit in Wheatstone-type bridges, but it gives no more than a first approximation to the condition of maximum sensitivity, because the optimum relationship between the impedances of the various arms depends on the impedances of the source and detector.

With reasonable precautions a bridge of this type can be used at frequencies up to a few tens of kHz. For the highest frequencies, bridge components of good quality are essential, as component calibrations become unreliable in the presence of stray reactance effects.

Errors can be significantly reduced by means of the following subsidiary experiment. The terminals of the inductor are connected together, and the bridge is rebalanced by adjustment of r and R. Suppose the value of R is now R'. Then strays are giving the effect of a fictitious inductance of magnitude

$$L' = C_0 R_0 R'$$

in series with the coil, and this should be subtracted from the apparent value of L.

The Owen bridge can also be used for the measurement of mutual inductance. We have seen in section 5.3.11 that if a transformer is connected as in Fig. 10.3, the effective self inductance is $L_1 + L_2 + 2M$. This is measured in an Owen bridge, and then the connections to one of the coils are reversed. The new self inductance is $L_1 + L_2 - 2M$, and this is also

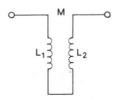

Fig. 10.3 **Circuit connections for measurement of mutual inductance**

measured. The difference between the measured values is $4M$. Why is the method unsatisfactory for a symmetrical transformer with a high coupling factor?

10.2.3   Schering bridge

Conceived in 1920 for the high-voltage measurement of dielectric loss at low frequencies, the Schering bridge has since proved useful for the measurement of capacitances as small as a few picofarads, and for operating frequencies as high as 1 MHz.

In the circuit of Fig. 10.4, series representation is used for the unknown imperfect capacitor, C being the constituent pure capacitance, and r a small series resistance which takes account of loss.

The general balance condition of section 10.2.1 is here best applied without modification, giving immediately

$$\left(r + \frac{1}{j\omega C} \right) j\omega C_0 = R \left(\frac{1}{R'} + j\omega C' \right).$$

Equating reals, we have after slight rearrangement

$$\frac{C}{C_0} = \frac{R'}{R}.$$

Equating imaginaries

$$\omega C_0 r = \omega C' R.$$

These are the two balance conditions for the bridge.

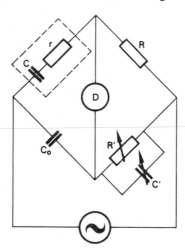

Fig. 10.4 Circuit of Schering bridge

Of the various bridge components, only C' and R' can provide independent satisfaction of the two balance conditions. As r is likely to be small, so also is C', and the latter often takes the form of a variable calibrated air condenser. For larger values a decade capacitance box may have to be provided. It is only the measurement of r which depends on the calibration of C', and this need not therefore be of high accuracy.

The principal interest is usually the measurement of C, and since C_0 appears in the first balance relation, the latter ought to be a good calibrated condenser. The dominant components are evidently C, C_0, R' and R, and the last two of these should be decade resistance boxes of good quality.

With series representation, the power factor of an imperfect capacitor is shown in section 6.11 to be approximately ωCr. Rearrangement of the two balance relations gives

$$\omega Cr = \omega C'R',$$

so that the bridge gives a direct measure of the power factor in terms of the values of the two variable components and the angular frequency of the source.

An important commercial application of the bridge is high-voltage insulation tests at power frequencies, on subjects ranging from cables to porcelain insulators. The capacitance C is often very small, in which case C_0 will also be small. If the values of R and R' are not large, the voltages which appear across the resistive arms of the bridge are quite small, and energy dissipation is insignificant. With the arrangement of Fig. 10.4, it is usual to earth the right-hand terminal of the source, and any risk of electrical shock involved in making adjustments to the variable components is then minimised. The large differences in impedance values between bridge arms greatly reduce the sensitivity of the bridge, although the high voltage provided by the source usually compensates adequately.

The bridge can be used for measurements over a wide range of capacitance and power factor. Measurements of small capacitance are in some respects facilitated by operation at high frequencies, because the impedance is correspondingly reduced. At the same time it is then necessary to select screening and earthing arrangements with greater care, a matter to which we shall give further attention in section 10.2.7.

10.2.4   Robinson frequency bridge

This bridge was devised in 1924 as a development suited for the measurement of frequency of a bridge due to Wien. The circuit appears in Fig. 10.5. The balance relation for the general Wheatstone-type network of

Fig. 10.1 can be written in the form

$$\frac{Z_4}{Z_1} = \frac{Z_3}{Z_2}.$$

This gives for the Robinson bridge

$$\left(R_4 + \frac{1}{j\omega C_4}\right)\left(\frac{1}{R_1} + j\omega C_1\right) = \frac{R_3}{R_2}.$$

Equating reals,

$$\frac{R_4}{R_1} + \frac{C_1}{C_4} = \frac{R_3}{R_2}.$$

Equating imaginaries,

$$\frac{1}{\omega C_4 R_1} = \omega C_1 R_4.$$

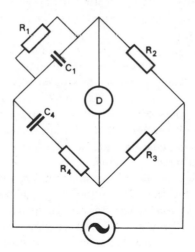

Fig. 10.5 Robinson frequency bridge

It is usual to employ equal fixed values for C_1 and C_4. Let

$$C_1 = C_4 = C, \quad \text{say.}$$

Also, let

$$R_1 = R_4 = R, \quad \text{say.}$$

Inspection of the two balance conditions shows that we shall require

$$R_3 = 2R_2,$$

and

$$\omega CR = 1.$$

Suitable fixed values in the ratio $1 : 2$ are selected for R_2 and R_3. Balance can be achieved at any frequency with R_1 and R_4 ganged and variable. The frequency is deduced from the relation

$$\omega = \frac{1}{CR}.$$

The bridge is suitable for use at frequencies up to several kHz.

10.2.5 Other Wheatstone-type bridges

Many additional Wheatstone-type bridges have been developed. A few of these are described briefly below.

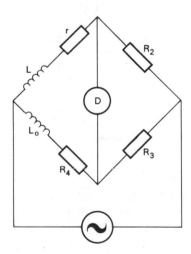

Fig. 10.6 Maxwell L/L₀ bridge

The Maxwell L/L_0 bridge of Fig. 10.6 appeared first in ballistic form, and was subsequently developed for use with an ac source by Wien in 1891. The balance conditions are

$$\frac{L}{L_0} = \frac{r}{R_4} = \frac{R_2}{R_3},$$

so that the bridge can be used for the comparison of self inductances. L_0 is preferably a calibrated, continuously variable inductor, and either R_2 or R_3

should also be variable. Care must be taken to avoid inductive coupling between the coils.

The Maxwell L/C bridge of Fig. 10.7 also appeared first in ballistic form, and was modified by Wien in 1891. The balance conditions are

$$\frac{L}{C} = R_1 R_3 = R_2 R_4.$$

This bridge has the advantage of dispensing with the calibrated variable self inductor, but requires instead a calibrated variable capacitor.

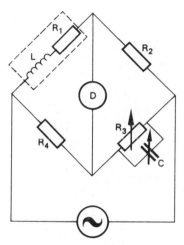

Fig. 10.7 Maxwell L/C bridge

In the Anderson L/C bridge of Fig. 10.8 the capacitor C is fixed, and balance is achieved by varying only resistances. The bridge was first introduced in ballistic form, and was modified for ac measurements by Rowland in 1898. The circuit is not of the true Wheatstone pattern, but the behaviour can be investigated fairly easily from first principles in terms of mesh currents. Alternatively the star-mesh transformation of section 11.6 is helpful. The balance conditions are

$$R_1 R_3 = R_2 R_4,$$

and

$$\frac{L}{C} = R_2 \left[r \left(1 + \frac{R_4}{R_3} \right) + R_4 \right].$$

The bridge can be balanced by adjustment of r and R_1. Notice that the circuit becomes indistinguishable from the Maxwell L/C bridge if r is reduced to zero.

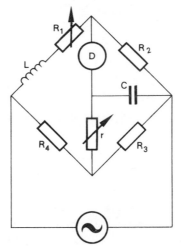

Fig. 10.8 Anderson bridge

The Carey—Foster M/C bridge of Fig. 10.9 was originally used for the ballistic measurement of mutual inductance. In its earliest form the circuit gave only aggregate balance, but in 1894 Heydweiller converted it so that a condition of absolute balance could be obtained, which would also be independent of frequency for ac conditions. This is again not a true Wheatstone-type bridge, and the behaviour is best analysed in terms of mesh currents. The balance conditions are

$$\frac{M}{C} = -R_2 R_4,$$

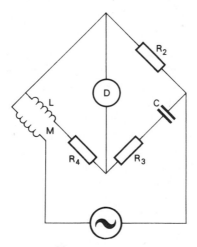

Fig. 10.9 Carey—Foster M/C bridge

and

$$\frac{L}{M} = -\left(1 + \frac{R_3}{R_2}\right).$$

Connections to the coils must be chosen to make the mutual inductance negative.

10.2.6   Layout

When assembling a Wheatstone-type bridge on the laboratory bench, it is helpful to group the components of the arms of the bridge in the same relative positions which they occupy in the circuit diagram. Many commercial components used in bridge measurements are supplied in rectangular boxes, with the connecting terminals positioned at one end of the box. In the arrangement of Fig. 10.10, the lengths of the connections between the bridge arms can be minimised, and there are no crossovers.

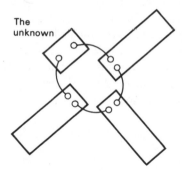

Fig. 10.10 Basic layout for Wheatstone-type bridge

The diagram is an over-simplification, because only one component is shown in each arm, and leads to the source and detector have been omitted. Notice that each component can be identified at a glance by relating its position to the circuit diagram. If an isolating transformer is used, it should be placed in the centre of the group, unless there is risk of inductive coupling to other components. The signal generator and detector should be placed behind the bridge components, and at a higher level. They are then easily seen, and their controls can be readily adjusted. Circuit faults are less difficult to identify and locate if the circuit layout is simple and logical.

10.2.7   Screening

Screening is of importance for all types of ac bridges. It will be discussed in some detail at this stage because of its special relevance for Wheatstone-type bridges to the Wagner earth, and to the use of the balanced and screened transformer, and ultimately to the obsolescence of the Wheatstone-type bridge.

It is possible for interaction to occur between electrical components even when they are not connected by any physical circuit. In rare cases the cause is conduction through imperfect insulation, but that effect is usually quite easily eliminated by suitably supporting the components. Magnetic pickup is more persistent. It can occur wherever two or more coils are present, and can cause direct interaction in bridge circuits between source and detector. Coupling between coils is greatly reduced if they are mounted with axes mutually perpendicular, and if the separation between them is made large. For low-frequency measurements, a high-permeability shield can provide efficient magnetic isolation. For high frequencies an enclosure made of copper sheet is preferred. This absorbs the available energy as eddy currents. The effective value of a component is altered by the provision of a screening device, so that this should be rigidly and permanently attached to the component if the measured value is to be always the same for a given frequency.

It is less easy to eliminate or compensate for stray capacitance. It can be associated with the plates and leads of a condenser, and with the individual turns of a coil. The most difficult aspect is its variability with the relative positioning of a circuit component and materials in the vicinity. The purpose of precautionary and corrective measures is reduction and stabilisation of the effect, and minimisation of any resulting error in measurement.

Even the stray capacitance between the two terminals of a component is likely to be variable. If, however, an enclosing metal box is provided, which is connected to one of the terminals, the stray capacitance between terminals is stabilised, and there remains only the problem of variable stray capacitance between the box and nearby conductors. If the box and the attached terminal can be earthed, this last effect is eliminated.

Where several components are connected in parallel and have one common terminal earthed, this same screening and earthing technique can be adopted for all. Each shield is connected by a separate lead to a common earth.

Where components are connected in series it is not possible to generalise, and each case must be given individual consideration. Suppose for example that one terminal of an individually shielded resistor-capacitor series combination is to be earthed, so that a choice can be made from the alternative arrangements of Figs. 10.11(*a*) and (*b*). In (*a*) the resistor would be shunted by the earth capacitance of the shield enclosing the

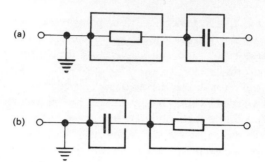

Fig. 10.11 Alternative arrangements for a shielded resistor-capacitor series combination

capacitor, and phase-shift would result. Alternatively (*b*) would be preferred because earth capacitance of the shield enclosing the resistor would simply directly shunt the capacitor. This stray capacitance could be stabilised by enclosing the shield in a second earthed shield. Double shielding is often favoured where components are connected in series, and for the situation just considered could take the form shown in Fig. 10.12.

A Wheatstone network with screened arms is illustrated in

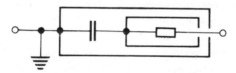

Fig. 10.12 Double shielding for resistor-capacitor combination

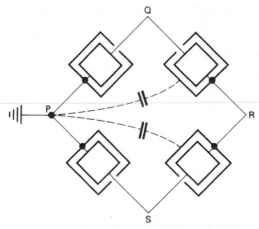

Fig. 10.13 Screening arrangements for components in arms of Wheatstone-type bridge

Fig. 10.13. The bridge is earthed at terminal P. For each of the two components attached to P, the screen and one appropriate terminal are earthed. No direct earthing is available for the components connected at R; the screens give rise to stable capacitance shunting each component, as well as stray capacitance between R and earth. The latter effect causes P and R to be linked by susceptance, which shunts either the source or detector, and does not affect the condition of balance.

10.2.8   Wagner earth

The screening arrangements which were discussed in the last section ensure stabilisation of stray capacitances associated with components in the bridge arms. Of course, when the detector is connected between Q and S in Fig. 10.13, additional earth capacitances are introduced which shunt the components in the bridge arms connected at P, and so disturb the balance condition. This effect was particularly inconvenient in the days when the detector was a headphone worn by the observer, because it varied with his position.

Errors due to this 'head effect' can be stabilised with the arrangement of Fig. 10.14. The earthing point for the bridge has been transferred

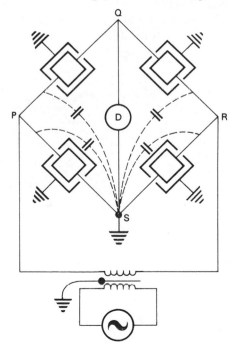

Fig. 10.14 Stray capacitance in Wheatstone bridge with earthed detector

to the detector connection at S, and the screens enclosing the various components in the bridge arms are now simply earthed. Earth capacitances associated with the source are minimised and stabilised by the use of an isolating transformer which is provided with an earthed screen of copper foil interposed between the primary and secondary windings. When the bridge is balanced, points Q and S are both at earth potential, so that earth capacitances associated with the detector carry no current.

The above procedure reduces and stabilises the magnitudes of the strays. These now take the form of capacitances-to-earth between components and earthed screens, and between the secondary winding and screen of the transformer. The strays disturb the balance condition by shunting the components connected in the bridge arms between P and S, and R and S.

Suppose now that 'dummy' bridge arms Z_1' and Z_2' are introduced into the Wheatstone network, as in Fig. 10.15. The earthing point is connected at the junction S' of Z_1' and Z_2'. The arrangement constitutes the Wagner earthing device. The detector is connected to S, and then the bridge is balanced in the usual way by adjustment of component values in the conventional bridge arms. The detector is next connected to S', and the bridge is balanced by adjustment of Z_1' and Z_2'. The characters and range of magnitudes of these components must be chosen so that a balance can be obtained in this way. If the detector connection is now transferred

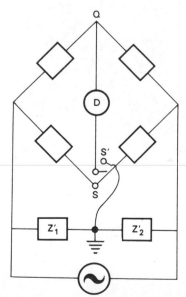

Fig. 10.15 Wheatstone-type bridge with Wagner earth modification

back to S, it will be found that the bridge is no longer balanced, because the changes in the dummy arms have altered the potential distribution relative to earth throughout the circuit. The conventional bridge arms are once more adjusted, and the process is continued until the bridge is balanced when connected both at S and S'. The various capacitances to earth now shunt only the dummy components Z_1' and Z_2', and make no contribution to the balance condition of the bridge.

The Wagner earthing device can also be adapted for use in some non-Wheatstone-type bridges.

10.2.9   Screened and balanced transformer

Impedance matching between source and bridge, and between bridge and detector, can be achieved by interposing transformers with suitable ratios of turns. This refinement gives improved power transference, and in consequence higher sensitivity. It is of course essential that there be no induction linkage between the two transformers, or between either transformer and any component in the bridge arms.

The primary and secondary coils should be separated by earthed copper screens. Stray capacitance between earth and the winding connected to the bridge will then be constant, so that the condition of balance is unaffected by the positioning or earthing of the source or detector. As viewed from the bridge arms, this stray capacitance is localized at the connections to the transformer. The bridge illustrated in Fig. 10.16 is earthed at junction P, and the small stray capacitances C_1 and C_2 due to the transformer shunt two of the arms of the bridge.

If a balance can be obtained with the ratio arms Z_1 and Z_2 identical in form, then the errors due to C_1 and C_2 can be eliminated as follows. The

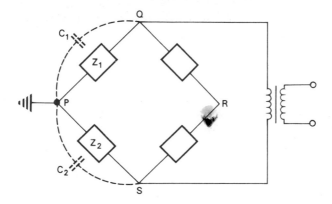

Fig. 10.16 Wheatstone-type network with transformer attached

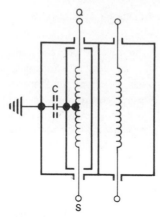

Fig. 10.17 Screened and balanced transformer

bridge is balanced first with the transformer connected as in the diagram, and then with the leads to the bridge interchanged. Average values of the component magnitudes obtained in the two measurements are substituted into the balance equations. This correction technique is unnecessary if C_1 and C_2 are equal.

For the transformer of Fig. 10.17, the primary and secondary coils are enclosed in separate earthed copper screens. One coil is additionally enclosed in an inner screen connected to the centre-tap. The symmetry of this construction enables the effective capacitances between the terminals of the doubly screened coil and the inner screen to be made equal. The device is called a screened and balanced transformer. As a result of careful design, the stray capacitances amount to only a few picofarads.

In the special case where the ratio arms of the bridge are identical, and the junction P in Fig. 10.16 is earthed, the centre-tap and inner screen of the transformer are at earth potential, and no current flows in the stray capacitor C.

10.2.10   *Limitations of Wheatstone-type bridges*

The Owen and Schering bridges are excellent for measurements at low frequencies, and alternative Wheatstone-type bridges can be used to advantage in special situations. A given bridge circuit can be used for measurements over a wide range of component values by adjustment of the ratio arms. The balance conditions of Wheatstone-type bridges are easy to derive, and the component relationships which are involved are simple and often independent of frequency.

But because two of the bridge arms are necessarily remote from the earth connection, there exist screening problems which become increasingly severe as the frequency is raised. The Wheatstone-type bridge is therefore seldom favoured for measurements at frequencies in excess of a few kHz.

10.3 Other bridges

In Wheatstone-type and related bridges, the state of balance involves equality of the potential differences developed across components in two networks connected in parallel across the source. The early 1940s saw the development of bridges utilising a new principle. Current is supplied to a detector via two routes. The detector current is reduced to zero at balance, the constituent currents being then equal in amplitude and opposite in phase. The end result is obviously no different so far as the detector is concerned, but novel forms of bridge emerge, facilitating simple and effective screening of components.

10.3.1 Transformer bridge

The ac source of Fig. 10.18 is connected via a detector across the complex admittances Y and Y', which are joined in parallel. Let us suppose that a condition is possible in which no current flows in the detector or source.

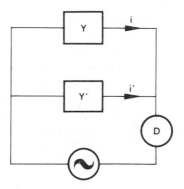

Fig. 10.18 Hypothetical resonant bridge circuit with infinite Q

Then if i and i' are the complex currents flowing in Y and Y' respectively,

$$i + i' = 0.$$

It follows that

$$Y + Y' = 0.$$

This is the condition of balance of the network. Evidently the susceptances and conductances of the two branches must separately total zero.

The first requirement presents no difficulty, but the second would necessitate the conductances being either both zero, or equal and of opposite sign. The implication is that parallel resonance occurs with infinite Q, and this is impossible in circuits containing only passive components.

We shall therefore turn our attention to two alternative modifications. In the first, one or each of the admittances is replaced by a T-network. This development will receive further consideration in section 10.3.2. In the other, the complex admittances are fed in antiphase by voltages derived from a transformer, the arrangement constituting a transformer bridge. The balance condition is now the physically realisable relation

$$Y = Y'.$$

The simplest form for the transformer bridge is that of Fig. 10.19. The two halves of the secondary winding are carefully constructed to provide emfs of equal amplitudes. The unknown admittance Y is to be compared with the variable calibrated admittance Y'. The earth connection is positioned so that one side of the detector and one side of each of the various admittances constituting Y and Y' are at earth potential. Each of these components can therefore be enclosed in simple earthed screens. At balance both terminals of the detector and the centre-tap of the transformer are at earth potential. The transformer should be fully screened to minimise and stabilise strays associated with the secondary coils and the primary circuit. Notice that at balance current continues to flow in the secondary coil and the two admittances. The term *bridge* can justifiably be used because in the absence of strays the balance condition is unaffected by interchange of source and detector.

The range of the bridge may be extended with the aid of tapped secondaries. If the emfs provided by the upper and lower sections of the secondary coil are in the ratio $n:1$, then the balance condition is

$$nY = Y'.$$

n can be chosen to be greater or less than unity. If sources of error such as ohmic resistance and leakage inductance can be made negligible by care in the design and construction of the transformer, n will be known very accurately in terms of the turns ratio, and is even independent of the currents in the two parts of the secondary. In normal circumstances the value of n is known to one part in 10^4, and in the transformer bridge used at the National Physical Laboratory for the absolute determination of the ohm (section 12.6.3), the error in n is considered to be less than one part in 10^7.

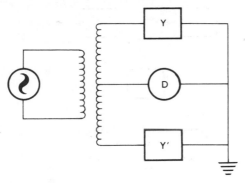

Fig. 10.19 Transformer bridge

For the measurement of pure capacitance, Y' could be simply a calibrated capacitor. Where measurement of conductance or negative susceptance is required, it is usual for Y' to be constructed as a network in which any continuous variable would be capacitive. A substitution technique may alternatively be favoured in which the unknown is introduced in parallel with a calibrated variable capacitor, so that direct compensation can be provided for the added susceptance. Provision to restore the conductance to the value for balance may have to be less direct.

10.3.2 *Bridged- and twin-*T *bridges*

These bridges were introduced by W. N. Tuttle, and take either of the two general forms illustrated in Fig. 10.20. The discussion which follows applies equally to either.

At balance there is no potential difference across the detector, and for purposes of analysis this can be replaced by a dead short. A T-network element of the bridge can therefore be represented with the source as in Fig. 10.21. The impedance presented by the network to the source is

$$Z_1 + \frac{Z_0 Z_2}{Z_0 + Z_2},$$

so that the current drawn from the source is

$$v \Big/ \left(Z_1 + \frac{Z_0 Z_2}{Z_0 + Z_2} \right).$$

It is left as a simple exercise to deduce that the current i delivered to the detector branch is given by

$$\frac{v}{i} = Z_1 + Z_2 + \frac{Z_1 Z_2}{Z_0}.$$

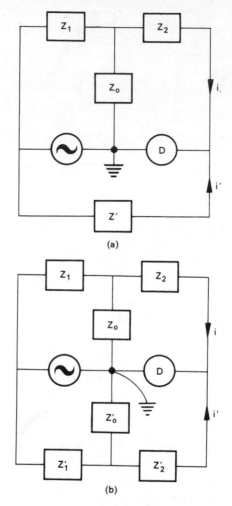

Fig. 10.20 (a) Bridged-T bridge and (b) twin-T bridge

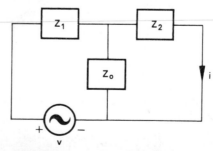

Fig. 10.21 T-network bridge element with source

The ratio $\dfrac{v}{i}$ is called the *transfer impedance* for the T-network. The reciprocal of this quantity is the *transfer admittance.*

In the complete bridge circuit current reaches the detector by two paths. For the complex branch currents which are indicated in Fig. 10.20 we have for balance

$$i + i' = 0.$$

If Y and Y' are the transfer admittances associated with these paths,

$$Y + Y' = 0.$$

It follows that if Z and Z' are the corresponding transfer impedances, then

$$Z + Z' = 0.$$

This is a convenient form for the balance conditions for this type of bridge.

The earth connection is at the common junction of the source and detector. If Z_0 embodies the unknown admittance, this also will then have one terminal earthed. The ability to earth the principal components in this way is a considerable advantage, enabling these bridges to be developed for use at frequencies ranging up to hundreds of MHz. It should be noted that not all the bridge components are directly earthed, and that both terminals of the components Z_1 and Z_1' are at all times at potentials remote from earth. Extreme care in screening is essential for all the bridge components at the high operating frequencies which can be used.

The absence of ratio arms sets a serious limitation on the range of admittances which can be measured with a given set of components. Moreover, the balance condition is inherently frequency-dependent, with obvious disadvantages, and is algebraically often so complicated that a substitution technique is the only practical method of use. Thus the impedance Z_0 may be made up as the unknown admittance connected in parallel with a calibrated variable admittance. Even so, it may be necessary to provide additional adjustment elsewhere in the bridge.

10.3.3 Tuttle-type bridges

A simple example of a bridged-T bridge appears in Fig. 10.22. This circuit can be used for the measurement of the self inductance L and series resistance r of the coil.

It was shown in the previous section that when a bridge of this general type is balanced, the sum of the transfer impedances of the two paths by which current reaches the detector is equal to zero. Thus

$$r + j\omega L + \frac{2}{j\omega C} + \frac{\left(-\dfrac{1}{\omega^2 C^2}\right)}{R} = 0.$$

Equating reals,

$$\omega^2 C^2 R r = 1,$$

and equating imaginaries,

$$\omega^2 L C = 2.$$

These relations give values for r and L respectively. Balance is obtained by adjustment of R and C. The capacitors are made as similar as possible and ganged.

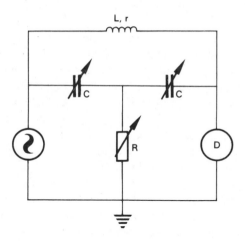

Fig. 10.22 Bridged-T bridge

For higher frequencies, the need for continuous variation of the resistor R sets a limit to the accuracy of measurement. There is also difficulty in the use of ganged variable capacitors where neither can be earthed at either terminal. It is unfortunate that the coil cannot be earthed, as this is the component under test. The potential of one end of the coil does of course tend to that of earth as balance is approached.

A second example of a bridged-T bridge is given in Fig. 10.23. The circuit differs from that of Fig. 10.22 only in that the resistor and coil have been interchanged.

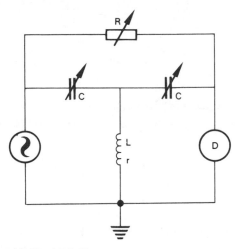

Fig. 10.23 A second bridged-T bridge

For balance, the sum of the transfer impedances is again zero, so that

$$R + \frac{2}{j\omega C} + \frac{\left(-\dfrac{1}{\omega^2 C^2}\right)}{r + j\omega L} = 0.$$

It is not difficult to separate out from this the relations

$$2\omega^2 LC = 1 - \omega^2 C^2 rR = \frac{4r}{R}.$$

There is now the advantage of being able to earth one side of the coil. It is again necessary to vary R and C for balance, and the variable resistor cannot now be satisfactorily earthed. In other respects the bridge possesses the disadvantages of the previous circuit, and the balance condition is algebraically rather more complicated.

A twin-T bridge is illustrated in Fig. 10.24. The balance relation is obtained by equating the sum of the transfer impedances of the two T-networks to zero. Thus

$$2R + j\omega C R^2 + \frac{2}{j\omega C'} + \frac{\left(-\dfrac{1}{\omega^2 C'^2}\right)}{R'} = 0.$$

Equating reals,

$$2\omega^2 C'^2 RR' = 1,$$

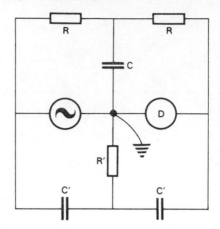

Fig. 10.24 Twin-T frequency bridge

and equating imaginaries,

$$\omega^2 CC'R^2 = 2.$$

The arrangement can be used for the measurement of frequency by setting $C = 2C'$ and $R = 2R'$. The two balance conditions are then indistinguishable, and the frequency is determined by the relation $\omega C'R = 1$. For measurement of high frequencies one might then choose to vary all three capacitors, with at least two of these being ganged together.

Another twin-T bridge is illustrated in Fig. 10.25. The coil is repre-

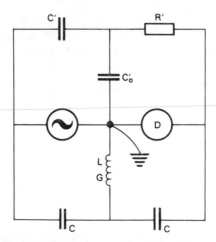

Fig. 10.25 Twin-T admittance bridge

sented as a self inductance L which is shunted by conductance G. The balance condition satisfies the equation

$$\frac{1}{j\omega C'} + R' + \frac{R'C_0'}{C'} + \frac{2}{j\omega C} + \left(-\frac{1}{\omega^2 C^2}\right)\left(G + \frac{1}{j\omega L}\right) = 0.$$

Separation of the reals and imaginaries gives, with a little rearrangement, the relations

$$G = \omega^2 C^2 R'\left(1 + \frac{C_0'}{C'}\right),$$

and

$$\frac{1}{\omega^2 LC} = 2 + \frac{C}{C'}.$$

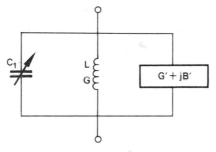

Fig. 10.26 **Detail of admittance bridge circuit**

The bridge could be used for the measurement of the electrical properties of the coil. The more usual arrangement is to provide L as a permanent bridge component shunted by a variable calibrated capacitor C_1, as detailed in Fig. 10.26. The net susceptance presented to the bridge at the coil connections must of course be suitably inductive at balance. The arrangement can be used to measure the unknown admittance $G' + jB'$, the technique being one of direct substitution. The bridge is first balanced in the absence of the unknown by adjustment of C_1 and C_0'. The unknown is now connected, and balance is restored by again adjusting C_1 and C_0'. The changes ΔC_1 and $\Delta C_0'$ in C_1 and C_0' obviously compensate for the additions of jB' and G' respectively, and the reader should not find it difficult to confirm that

$$B' = -\omega\,\Delta C_1,$$

and

$$G' = \frac{\omega^2 C^2 R'}{C'}\,\Delta C_0'.$$

10.4   Practical considerations

10.4.1   Sources and detectors

For ac bridges generally, detection equipment should be proceded by a stage of adjustable electronic amplification, so that variable sensitivity is available. The input impedance of electronic amplifiers is customarily high, and may be as much as several megohms. Source impedances are usually relatively low, a few hundred ohms being typical. Greater sensitivity is therefore likely to be obtained if the source and detector connections are selected in such a way that the bridge presents the latter with the greater internal impedance.

During the course of bridge measurements, the output of the detector amplifier should preferably be displayed continuously by means of a cathode-ray oscilloscope. A constant check can then be kept on the waveform of the voltage appearing at the detector terminals of the bridge. It is important to be able to confirm that this is reasonably free of overtones and mains hum, and of signals picked up from neighbouring apparatus. These unwanted constituents are likely to be present in greatest relative strength in the vicinity of balance. This is because the balance condition is always frequency-dependent, either because of the nature of the bridge circuit, or because of strays associated with the individual components. By watching the trace on the cathode-ray tube, the operator can decide whether he is approaching balance, and whether the overall sensitivity is limited to a significant extent by the masking effects of unwanted signals. He may be content to make the final adjustments while observing the trace, but much greater sensitivity is obtained by rectifying the signal and displaying the resulting unidirectional current as the deflection of a moving-coil instrument. Filtering is essential if the signal/noise ratio at the detector is low enough to cause intolerable impairment of sensitivity.

The overtone content of the signal presented to the detector usually originates in the source. It may alternatively be created by a nonlinear bridge component, as for example in transistor measurements, or where an iron-cored coil is present. The balance will now depend on the magnitude of the alternating current flowing in the component, and the signal should be reduced to the lowest level providing adequate sensitivity, so that an approximation to linear behaviour is obtained.

A superheterodyne receiver is used for amplification and detection of frequencies in excess of a few MHz. It is convenient to amplitude-modulate the source at a frequency of the order of 1 kHz. The modulation is extracted by the receiver and displayed by means of a cathode-ray oscilloscope or moving-coil rectifier instrument.

Stability of the source frequency is obviously desirable, especially if the balance condition of the bridge is frequency-dependent. For detectors with limited bandwidth, such as the superheterodyne receiver, any drift in signal frequency produces large variations in sensitivity.

Additional reading

BALDWIN, C. T., *Fundamentals of Electrical Measurements.* Harrap, 1961.

BROOKES, A. M. P., *Advanced Electric Circuits.* Pergamon, 1966.

BUCKINGHAM, H. and PRICE, E. M., *Principles of Electrical Measurements.* English Universities Press, 1967.

HAGUE, B. (and FOORD, T. R.), *Alternating Current Bridge Methods.* Pitman, 1971.

KARO, D., *Electrical Measurements.* Part II. Macdonald, 1953.

TURNER, R. P., *Bridges and other Null Devices.* Foulsham—Sams, 1968.

Examples

1. The four arms in an Owen bridge taken in cyclic order contain components (*a*) R_1, (*b*) C_0, (*c*) R_2 and C, and (*d*) an inductor L possessing some series resistance r_0, which has a resistor r connected in series with it. The operating angular frequency is 5000 rad s^{-1}. A preliminary balance is obtained with $C_0 = 0.05 \mu F$, $r = 170 \Omega$, $C = 0.1 \mu F$, $R_1 = 400 \Omega$, and $R_2 = 5 k\Omega$. Give values for L and r_0.

    Suggest approximate optimum values for R_1, C_0 and R_2, assuming the impedances of source and detector are approximately equal.

    If r were intended to have a value of about 20 Ω for balance, what would be a suitable value for C when R_1 and C_0 are optimised?

2. The four arms in a Schering bridge taken in cyclic order contain components (*a*) R, (*b*) R′ and C′ connected in parallel with each other, (*c*) C_0, and (*d*) the unknown capacitor C possessing some series resistance r. Balance occurs for $R = R' = 1 k\Omega$, $C' = 100$ pF, and $C_0 = 0.1 \mu F$. Determine the values of C and r. For what approximate angular frequency is the bridge most sensitive, assuming the impedances of the source and detector are roughly equal?

3. The capacitance and power factor of a 100 pF commercial capacitor are to be measured in a Schering bridge at an angular frequency of 10^7 rad s^{-1}. A standard 1000 pF capacitor and a standard 1 kΩ

resistor are available as bridge components. The power factor is thought to be about 0.05. Give the probable approximate values at balance of the two variable components in the bridge. If the actual values of these at balance are 125 Ω and 62.5 pF, what are the effective series capacitance and resistance and the approximate power factor of the commercial capacitor?

4. In the bridged-T bridge of Fig. 10.22, $C = 12.5$ pF and $R = 16$ kΩ for balance, the operating frequency being 100 MHz. Obtain values for L and r.

5. In the twin-T bridge of Fig. 10.24, component values for balance are $C = 2C' = 20$ pF, and $R = 2R' = 1$ kΩ. Determine the angular frequency of the source.

6. Show analytically that if the standard capacitor in an Owen bridge develops shunt leakage resistance it will still be possible to balance the bridge in the usual way.

Circuit theorems

Kirchhoff's two laws for electric circuits (sections 2.1.3, 2.5) embody the laws of conservation of charge and energy in forms applicable directly to circuit analysis. If the voltage—current relationship for each component in a given circuit is known, they enable detailed characteristics of the complete assembly to be predicted. The analysis may be very time consuming for circuits which are not especially simple in structure.

A circuit theorem is a statement of a relationship developed from the basic laws of circuit behaviour, which provides a more advanced starting point from which analysis may proceed. Some circuit theorems are very obvious and scarcely require formal proof, and rigorous proofs can in any case be unpleasantly tedious. The intention here is to discuss some of the more common theorems, and to give illustrations of their use. Those proofs which are given will be of an elementary nature.

It is quite usual to state circuit theorems in terms of direct current behaviour. Saving the power-transfer theorem which we have already encountered in section 6.12, the extension to alternating current conditions is usually obvious, and presents little difficulty.

11.1 Superposition Theorem

The Superposition Theorem states that *the current in any branch of a network is the sum of the currents in that branch contributed by each source of emf in the network acting alone.*

The theorem is an example of a linear superposition principle, and as such is sufficiently obvious not to require justification. It is only valid if the individual circuit components are linear in behaviour, and provided all other sources are replaced by their internal impedances when the current due to any one is being calculated.

Care is necessary where two or more ac sources are present, because

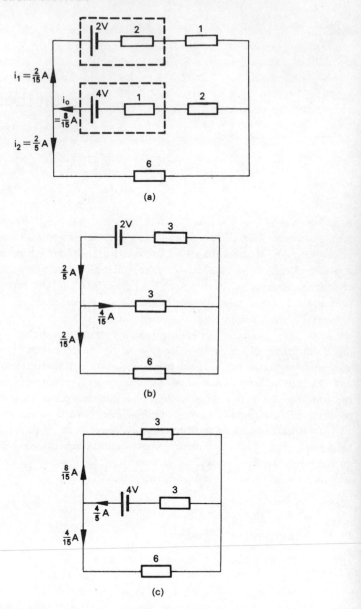

Fig. 11.1 (a) The given network, (b) current distribution in absence of emf of 4 V battery, (c) current distribution in absence of emf of 2 V battery

phase-differences between constituent currents in a given branch will depend in part on phase-differences between sources.

As an illustration of the theorem, consider the dc circuit of

Fig. 11.1(*a*), in which all resistances are in ohms. If the steady current distribution maintained by the two batteries is investigated by, for example, the mesh current technique, the branch currents i_0, i_1 and i_2 are found to have the values indicated in the diagram. Let us now instead use the Superposition Theorem. If the emf of the 4 V battery is reduced to zero, the circuit takes the appearance of Fig. 11.1(*b*). The effective resistance of the 3 and 6 ohm resistors in parallel is 2 ohms, so that the load presented to the 2 V battery is 3 + 2 ohms. It is easy to see that the current distribution is therefore that shown in the diagram. Let it now be the emf of the 2 V battery which is reduced to zero. The load presented to the 4 V battery is 3 + 2 ohms, giving in consequence the current distribution of Fig. 11.1(*c*). It can now be seen that the current totals in each of the branches agree with the values given in Fig. 11.1(*a*).

It happens that the theorem does not really simplify this particular circuit problem. It is however especially useful in investigating the effects of inserting additional emfs in networks, and it is also an essential basis for other theorems.

11.2   Thévenin's Theorem

This may be stated: *If i is the current which flows in a short-circuit connected between two terminals in a network, and v is the potential difference which appears between the terminals when the short-circuit is removed, then if the magnitudes of all emfs within the network are reduced to zero, the impedance presented at the terminals to an external source of emf is* v/i.

The implication is that the properties of a linear network as observed

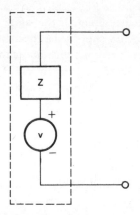

Fig. 11.2 Equivalent circuit of linear network

at a pair of its terminals are indistinguishable from those of a source of emf (v) with internal impedance (Z), so that it can be represented as in Fig. 11.2. This is almost self-evident, and might be preferred as a statement of the theorem.

Thus an amplifier or signal generator appears as a source of emf in series with an internal impedance, when viewed from its output terminals. An important consequence is that the internal impedance can be measured with the aid of an external source of emf, with the internal emf of the network reduced to zero. The reader may notice that the existence and validity of the theorem have been more than once tacitly assumed in earlier chapters.

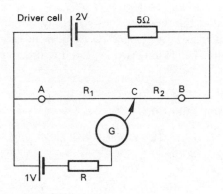

Fig. 11.3 Calibrated metre bridge for measurement of emf

The theorem will now be used to calculate the off-balance current in the dc potentiometer circuit of Fig. 11.3. The emf of the 1 V cell is determined by locating the balance point of the slider on the uniform 15 ohm bridge wire AB, which we may suppose has been calibrated with the aid of a standard cell. R is a safety resistor which limits the galvanometer current. It is obvious that at balance $R_1 = 10$ ohm and $R_2 = 5$ ohm. To fix ideas, suppose we are required to determine the galvanometer current when the slider is moved 1/1000 of the length of AB away from balance. Now the potential difference across the full length of the bridge wire is

$$\frac{R_1 + R_2}{R_1} \cdot 1 \text{ V} = 1\frac{1}{2} \text{ V}.$$

The potential difference provided by the driver cell across AC will therefore be either $\left(1 + \dfrac{1\frac{1}{2}}{1000}\right)$V or $\left(1 - \dfrac{1\frac{1}{2}}{1000}\right)$V, depending on whether the

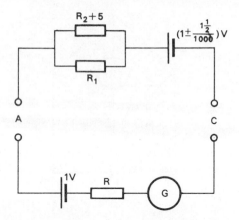

Fig. 11.4 Equivalent circuit of metre bridge, with galvanometer branch

tapping point is moved to the right or left of the balance position. In view of Thévenin's Theorem, the equivalent circuit will be that of Fig. 11.4. The galvanometer circuit is unchanged, but the potentiometer circuit is now represented as a source of emf equal to the open-circuit voltage across AC when the galvanometer connection is absent, in series with an internal resistance equal to the input resistance at AC when the driver cell is replaced by a short-circuit. We can use as a good approximation the values of R_1 and R_2 appropriate for balance, so that this internal resistance is closely equal to 5 ohms. The off-balance current which flows when the galvanometer circuit is reconnected is therefore

$$\pm\frac{1\frac{1}{2}}{1000}\bigg/(5 + R + G)\ \text{A} = \frac{3}{2(5 + R + G)}\ \text{mA}.$$

Thévenin's Theorem can also be used to reduce the labour involved in calculating the off-balance current in the Wheatstone bridge of Fig. 11.5. For simplicity the battery will be supposed to lack internal resistance. We shall require to find the equivalent circuit of the bridge as viewed from the galvanometer terminals. The internal emf is the potential difference which develops at these terminals when the galvanometer branch is open-circuited, and this is equal to the difference v between the potential differences across Q and R. Thus

$$v = v_0\left(\frac{Q}{P + Q} \sim \frac{R}{R + S}\right).$$

Now replace the bridge battery by a short circuit. The equivalent circuit of

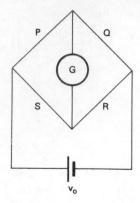

Fig. 11.5 Unbalanced Wheatstone bridge

the bridge takes the form of Fig. 11.6, so that the resistance presented at the galvanometer connections is

$$1 \Big/ \left(\frac{1}{P} + \frac{1}{Q} \right) + 1 \Big/ \left(\frac{1}{R} + \frac{1}{S} \right) = r, \quad \text{say.}$$

The off-balance current is $\dfrac{v}{r + G}$.

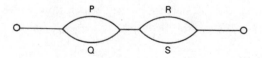

Fig. 11.6 Equivalent circuit of internal resistance of bridge

11.3 Norton's Theorem

It is sometimes convenient to employ Norton's Theorem in place of Thévenin's Theorem, to which it is equivalent. Norton's Theorem states that *a linear network may be represented at any pair of its terminals as a current source shunted by an impedance,* in the manner of Fig. 11.7. The current i flows through the output terminals if they are connected directly together. The value of the impedance Z can be identified by equating to the product Zi the voltage developed across the terminals in the absence of an external load.

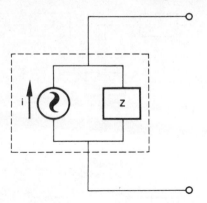

Fig. 11.7 Equivalent Norton generator

11.4   Reciprocity Theorem

The Reciprocity Theorem states that *if a source of constant emf and zero internal impedance which is inserted in a branch AB of a passive linear network produces a certain current in a branch CD, then the same source in branch CD produces the same current in branch AB.*

The proof of the theorem is as follows. The T-section of Fig. 11.8(*a*) is a simple representation of a four-terminal passive linear network. Let a source of constant emf *v* and zero internal impedance be inserted between terminals A and B, as in Fig. 11.8(*b*), and suppose a current *i* flows in a short-circuit linking terminals C and D. The quantity *v/i* is Tuttle's transfer impedance (section 10.3.2) for the T-network, and is related to the circuit component values according to

$$\frac{v}{i} = Z_1 + Z_2 + \frac{Z_1 Z_2}{Z_0}.$$

This equation is seen to be unaltered by interchange of Z_1 and Z_2, a modification equivalent to interchange of the branches containing the source of emf and the branch current *i*. The theorem is therefore verified.

If no current flows in one branch of a passive network when a source of emf is inserted in another branch, then the two branches are said to be *conjugate.* For example, in a balanced Wheatstone bridge the source and detector branches are conjugate.

The Reciprocity Theorem is inapplicable if any impedance is transferred with the source of emf, because although the summation $Z_1 + Z_2$ remains unchanged, the product $Z_1 Z_2$ in general does not. Consider for example the effect of interchange of source and detector in any bridge

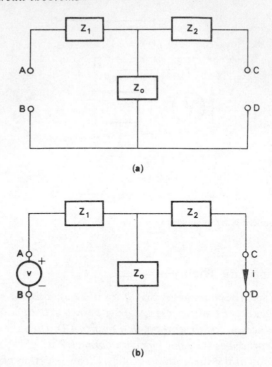

Fig. 11.8 Equivalent four-terminal network

circuit; the overall sensitivity will evidently be unchanged only if the impedances of the two happen to be equal, in which special situation the effect is as if no impedance is transferred.

11.5   Compensation Theorem

This states that *the changes in the current distribution which occur in a linear network when an impedance Z is inserted in a branch carrying current i are the same as when a source of emf iZ and zero internal impedance is inserted in the same branch in opposition to the current.*

Notice that i is the value of the *unmodified current* in the branch. If the impedance Z and the source of emf are inserted simultaneously, but with the terminals of the latter interchanged so that it provides support for the flow of current i, then the potential drop iZ in the impedance and the additional emf $-iZ$ total zero. The combination will therefore produce no change anywhere in the network. But, according to the Superposition Theorem, the impedance and the emf contribute independent changes to the distribution, so that their effects in any branch must be supposed to be equal and opposite in sign. Their separate effects will be identical if the terminals of the source of emf are once more interchanged, so that the Compensation Theorem is verified.

The Compensation Theorem enables the effects of an impedance change to be determined by considering instead the influence of an additional emf. Consider for example the balanced Wheatstone bridge of Fig. 11.9. The response to a given change in the resistance of one arm of the bridge can be investigated as follows. The current i_1 is first calculated.

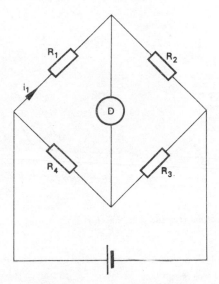

Fig. 11.9 Balanced Wheatstone bridge

Now let R_1 increase by an amount ΔR_1. According to the theorem, balance can be restored with the aid of an additional emf $i_1 \Delta R_1$ which is connected in series with R_1 in support of the flow of current i_1. One need therefore merely derive a formula for the current in the galvanometer due to this emf. The calculation is especially simple if the battery branch is entirely lacking in resistance, in which case it can be replaced for purposes of analysis by a short-circuit.

11.6 Star-mesh transformation

The star-mesh transformation is known alternatively as the Y-delta or star-delta transformation. It gives the conditions for equivalence of the three-terminal networks illustrated in Fig. 11.10.

11.6.1   *Star elements in terms of delta elements*

In order to find values for Z_A, Z_B and Z_C in terms of Z'_A, Z'_B and Z'_C, we can study the special case where the three terminals are unconnected to any external circuit. If the networks are equivalent in all respects, the relations between the component values cannot be affected by the attachment of additional impedances. The input impedance between A and B is

$$Z_A + Z_B = Z'_C \frac{Z'_A + Z'_B}{\Sigma Z'},$$

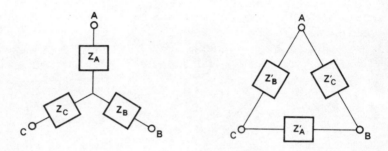

Fig. 11.10 Three-terminal star and delta networks

where

$$\Sigma Z' = Z'_A + Z'_B + Z'_C.$$

Consideration in turn of the impedances at the pairs of terminals B, C and C, A gives the additional relations

$$Z_B + Z_C = Z'_A \frac{Z'_B + Z'_C}{\Sigma Z'},$$

and

$$Z_C + Z_A = Z'_B \frac{Z'_C + Z'_A}{\Sigma Z'}.$$

A formula for Z_A in terms of the components of the delta configuration is obtained by adding the first and last of these equations together, and subtracting the last-but-one, viz:

$$2Z_A = \frac{2Z'_B Z'_C}{\Sigma Z'}.$$

Therefore

$$Z_A = \frac{Z'_B Z'_C}{Z'_A + Z'_B + Z'_C}.$$

The form of the corresponding relations for the components Z_B and Z_C is obvious.

11.6.2 Delta elements in terms of star elements

The various components of Fig. 11.10 will now be represented as admittances. For example, the symbol Y_A replaces $1/Z_A$. We shall again study the special case in which the three terminals are unconnected via any external circuit, and in addition terminals B and C will be supposed to be connected directly together by a short-circuit. If the networks are equivalent in all respects, the relations between the component values cannot be affected by the attachment of additional admittances. The input admittance between terminals A and B will now be

$$Y'_B + Y'_C = 1 \Big/ \left(\frac{1}{Y_A} + \frac{1}{Y_B + Y_C} \right).$$

Using the substitution

$$\Sigma Y = Y_A + Y_B + Y_C,$$

this reduces to

$$Y'_B + Y'_C = Y_A \frac{Y_B + Y_C}{\Sigma Y}.$$

Corresponding relations are given below for the input admittances between B, C and C, A, the short-circuit being connected successively in parallel with Y'_B and Y'_C. Thus

$$Y'_C + Y'_A = Y_B \frac{Y_C + Y_A}{\Sigma Y},$$

and

$$Y'_A + Y'_B = Y_C \frac{Y_A + Y_B}{\Sigma Y}.$$

If the last two of these three relations are added together, and the first is subtracted, the following equation emerges:

$$Y'_A = \frac{Y_B Y_C}{Y_A + Y_B + Y_C}.$$

Similar formulae can be written down for Y'_B and Y'_C. Notice the resemblance to the transformations of section 11.6.1, with the difference that admittances take the place of impedances.

11.6.3   *Realisability of equivalent network*

For a given initial circuit configuration, it may not be possible to construct an equivalent network with physically available components. This does not mean that the transformation employed is invalid, or that it is any less helpful in analysis. If there are pressing reasons for the actual construction of the equivalent network, it may be possible to select a combination of components which has the desired characteristics at a single frequency. If, for example, negative resistance is required, an active device may be available which behaves suitably over the desired range of conditions.

11.6.4   *Applications*

The theorem can be extended in modified form to the general case of n-terminals, and the possible range of applications is then very wide. For example, the four-terminal version can be used to express the stray earth capacitances at the vertices of a Wheatstone net as shunting capacitances in parallel with each bridge arm, and with the source and detector.

We shall restrict our interest here to the three-terminal form of the transformation, and will consider as a first illustration the calculation of the galvanometer current i_G in the unbalanced Wheatstone bridge circuit of Fig. 11.11(a). The star-mesh transformation for components connected directly between terminals A, B and C reduces this to the form (b), and the circuit can then be further simplified to the form (c). The new component values are of course all calculable in terms of the originals, so that the battery current i_0 can be obtained from the relation

$$\frac{v_0}{i_0} = R_B + 1 \left/ \left(\frac{1}{1/R_1 + 1/R_2} \right) \right. .$$

We next obtain expressions for the currents i_P and i_S. Thus

$$\frac{i_0}{i_P} = \frac{P + S + R_A + R_C}{S + R_C},$$

and

$$\frac{i_0}{i_S} = \frac{P + S + R_A + R_C}{P + R_A}.$$

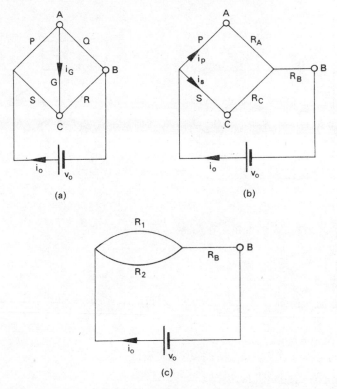

Fig. 11.11 (*a*) **Unbalanced Wheatstone bridge**, (*b*) **star conversion**, (*c*) **equivalent circuit**

Then further

$$\text{pd across G} = \text{pd across S} - \text{pd across P}$$

$$= S i_S - P i_P$$

$$= G i_G.$$

A complete expression for the galvanometer current i_G is easily obtained from the above relations.

As a second illustration we shall obtain the balance conditions of the bridged-T network of Fig. 11.12(*a*). With the aid of a star-delta transformation, the star connecting terminals A, B and C becomes the delta of diagram (*b*). The admittances Y'_A and Y'_B do not affect the balance condition as they shunt the detector and source respectively. Now at balance, the combined admittances of the coil and Y'_C total zero, so that

$$\frac{1}{r + j\omega L} + \frac{(j\omega C_0)^2}{2j\omega C_0 + 1/R} = 0,$$

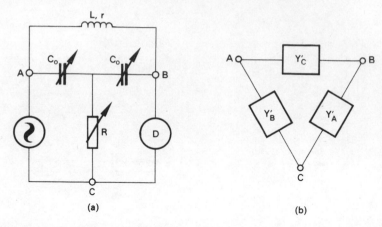

Fig. 11.12 (*a*) Bridged-T bridge, (*b*) delta network

or

$$2j\omega C_0 + \frac{1}{R} = \omega^2 C_0^2 (r + j\omega L).$$

Equating the reals,

$$\omega^2 C_0^2 rR = 1,$$

and equating the imaginaries,

$$\omega^2 LC_0 = 2.$$

These are the conditions of balance of the bridge.

Additional reading

BROOKES, A. M. P., *Advanced Electric Circuits.* Pergamon, 1966.

DUFFIN, W. J., *Electricity and Magnetism.* McGraw-Hill, 1965.

EDMINSTER, J. A., *Electric Circuits.* Schaum's Outline Series, 1965.

FAULKNER, E. A., *Principles of Linear Circuits.* Chapman and Hall, 1966.

HARNWELL, G. P., *Principles of Electricity and Electromagnetism.* McGraw-Hill, 1949.

MEADOWS, R. G., *Electric Network Analysis.* Athlone Press, 1972.

SCANLAN, J. O. and LEVY, R., *Circuit Theory.* Vol. I. Oliver and Boyd, 1970.

SHIRE, E. S., *Classical Electricity and Magnetism.* Cambridge University Press, 1960.

STARR, A. T., *Electric Circuits and Wave Filters.* Pitman, 1938.

Examples

1. The four arms of a Wheatstone bridge taken in cyclic order have the resistances 1000, 1000, 985 and 1015 ohms. A galvanometer provided with a safety resistor giving a total series resistance of 500 Ω connects the junction of the two equal resistors to the junction of the two unequal resistors. The 2 V battery is of negligible internal resistance. Obtain an approximate value for the current which flows in the galvanometer.

2. A 3 V battery of negligible internal resistance, a 12 Ω resistor and a 24 Ω slide wire of length 5 m are connected as a closed series circuit. The slide wire is standardised in the usual way, using a Weston cell and a moving coil galvanometer with a series safety resistor. What largest approximate value can be tolerated for the total series resistance of the standardising circuit, if a minimum current of 0.5 μA can be detected in the galvanometer, and it is desired to measure the balance length to within ±0.5 mm? The calculation is simplified without significant loss of accuracy if the emf of the Weston cell is taken to be 1 V.

3. A standardised 4 Ω bridge wire which is connected to the terminals of a 2 V battery of negligible internal resistance is to be used to calibrate an ammeter. The ammeter is connected in a separate closed series circuit with a standard 2 Ω resistor, a 2 V cell and a variable resistor. The potential difference across the standard resistor is balanced against the potential difference developed across a length of the bridge wire, using a galvanometer of resistance 48 Ω to detect the balance. If the smallest galvanometer current which can be detected is 0.05 mA, with what percentage accuracy can the balance length be identified when the current in the ammeter circuit is 0.5 A?

4. A 7 V battery having an internal resistance of 20 Ω is connected in parallel with a 14 V battery having an internal resistance of 80 Ω, terminals of unlike polarity being connected together. A 40 Ω resistor is connected as an external load. Use the Superposition Theorem to determine the currents flowing in the batteries.

12

Electrical units

12.1 Introduction

In 1948 the 9th General Conference of Weights and Measures (CGPM) recommended adoption of the MKSA system of units, in which definitions of the various electrical quantities are related to four basic units, namely the metre, kilogram, second and ampere. At the 11th CGPM in 1960, the system was extended to form part of the Système International d'Unités (SI), which at the present day has gained general acceptance internationally, and is employed almost universally in electrical literature.

During the previous 100 years a remarkable variety of systems had been in use at one time or another, with some inevitable confusion. There were even instances of the same name being attached to differing systems. Relative merits have been debated for decades, with an intensity which can surprise the uncommitted, who may understandably be reluctant to accept that any one system might possess a monopoly, or even a majority, of virtues.

We are concerned here with the behaviour of linear circuits, and our involvement with the SI system is rather limited. But a high proportion of the measurements required to establish and maintain SI as a working system are developed around linear circuits, and fall therefore within our proper area of interest. With the continuing development of new materials and methods, the present situation is merely an evolutionary phase. A flexible attitude must be maintained concerning the techniques whereby the various units are established, and even towards the underlying definitions.

12.2 Fundamental mechanical units

The metric system is based on the metre as the unit of length. The metre was originally intended to be one ten-millionth part of a meridional quadrant of the Earth, but was eventually redefined simply as the distance between two engraved lines on a certain platinum-iridium bar preserved in the International Bureau of Weights and Measures at Sèvres, near Paris. Since 1960 the metre has been defined as the length occupied by 1 650 763.73 wavelengths *in vacuum* of the radiation corresponding to the transition between the levels $2p_{10}$ and $5d_5$ of the krypton-86 atom.

The unit of mass in the metric system is the kilogram. This was originally defined as the mass of a cubic decimetre of water at its temperature of maximum density (4 °C). Since 1889 an international prototype kilogram has been maintained at Sèvres in the form of a cylinder of platinum-iridium alloy.

The second of time is intended to approximate closely to 1/86 400 of the mean solar day. Because of variations in the rotation of the Earth, a unit of time which is defined in this way tends to vary in the course of any year, and to undergo irregular longer-term changes. The variations are extremely small, but their existence is clearly undesirable in a basic unit. Definitions based on alternative astronomical observations have also been used, but there are obvious benefits to be gained from the establishment of a standard of time related to purely laboratory-based observations. The second of time is now defined as the duration of 9 192 631 770 periods of the radiation corresponding to the transition between the two hyperfine levels of the ground-state of the caesium-133 atom.

Additional mechanical units are employed for quantities such as force and energy. These require no fundamentally new definitions, as they are based entirely on the already established units of mass, length and time. Thus the newton (N), which is the unit of force, is the force needed to accelerate a mass of 1 kg at the rate of 1 ms^{-2}, and the joule (J), which is the unit of energy, is the work done by a force of 1 N acting over a distance of 1 m measured in the direction of the force.

12.3   Systems of electrical units

12.3.1   Basic CGS systems

The two early centimetre-gram-second (CGS) systems found wide acceptance over a long period of time, and it is from these that the various more recent systems are derived.

The CGS electrostatic system of units (CGS ESU) was based on Coulomb's Law. The results of experiment are consistent with the view that the force F between point charges of magnitudes q_1 and q_2 depends on the distance r between them according to the relation

$$F \propto \frac{q_1 q_2}{\epsilon r^2}.$$

Here the value of ϵ depends on the nature of the medium, and in this context is conveniently defined to be unity for vacuum. If the constant of proportionality is also taken as unity, the relation enables a unit of charge to be defined which exerts unit force of repulsion on a similar charge placed unit distance away *in vacuum*. With the force F and distance r measured in centimetre-gram-second units, the resulting CGS unit of charge forms the basis of the CGS ESU system.

Experiment suggests that the force F per unit length between two infinitely long straight parallel conductors should depend on the strengths of the steady currents i_1 and i_2 flowing in them according to the relation

$$F \propto \frac{\mu i_1 i_2}{a},$$

where a is the distance between the conductors. The value of μ depends on the nature of the medium, and can be defined as unity for vacuum. If the constant of proportionality is defined to be two, then the formula can be used to define a unit of current which, when flowing in conductors arranged as above in vacuum, gives rise to a force between them of 2 dynes per centimetre length of conductor when the separation is 1 centimetre. This unit of current could be taken as the basis of the centimetre-gram-second electromagnetic system of units (CGS EMU).

The CGS ESU and CGS EMU systems are seen to derive from definitions of units of charge and current respectively. The relation

$$q = \int_{t_1}^{t_2} i \, dt$$

enables a unit of charge to be defined in terms of the CGS EMU of current. There is no reason why this should correspond with the CGS ESU of charge. The ratio

$$\frac{\text{CGS EMU of charge}}{\text{CGS ESU of charge}}$$

is in fact very large indeed, and was shown theoretically by Maxwell to be numerically equal to the speed of light in empty space in CGS units.

12.3.2   Practical and international units

The CGS EMU system was first proposed by Weber as a basis for a system of electrical units related to mechanical units. The magnitudes of the more common electrical units established in this way were, however, not well suited for the principal applications of the time. Within a few years a committee appointed by the British Association for the Advancement of Science, and under the chairmanship of Maxwell, had recommended in 1861 the adoption of modifications which found immediate acceptance in this country, and which determine the sizes of present-day electrical units.

A principal recommendation by the committee was the introduction of a practical unit of resistance to be called the ohm. This was to be equal to 10^9 times the absurdly small resistance unit of the CGS EMU system. At the same time it was decided that a practical unit of potential difference should be defined, equal to 10^8 CGS EMU, and called the volt. This unit is comparable with the emfs of primary cells. These changes necessitated in turn the introduction of a new unit of current, for consistency in the relation

$$\text{resistance} = \frac{\text{voltage}}{\text{current}}.$$

This unit was called the ampere, and was equal to 10^{-1} of the CGS EMU of current. The new electrical units gained formal international acceptance in 1881.

Because of the need to ensure general uniformity of standards, reproducible and portable practical standards had subsequently to be developed. Thus a 'mercury ohm' was defined, in the form of a column of mercury with specified dimensions, temperature, and electrode geometry. The corresponding unit adopted for current was the 'silver ampere', which was the steady current depositing silver electrolytically from silver nitrate, at a specified rate and under carefully detailed conditions.

At the London International Congress on Electrical Units and Standards which was held in 1908, it was acknowledged that increasing precision in electrical measurements was revealing significant differences between the reproducible and theoretical electrical standards. The former were therefore designated International Electrical Units, with a view to emphasising the distinction, and it is evident that their usefulness was already declining, as direct absolute measurements could now be made with greater accuracy than the practical standards could be constructed. The International Electrical Units were in fact finally abandoned in 1948.

12.3.3   Giorgi system

We encountered in the previous section the factors 10^8 and 10^{-1} which are involved in the relation between CGS EMU and the practical units of voltage and current respectively. The energy delivered into a circuit is proportional to the product of voltage and current, and the practical unit of electrical energy (joule) is therefore 10^7 times as great as the CGS unit (erg). It was proposed by Giorgi in 1902 that this ratio could be conveniently absorbed if the centimetre and gram were displaced from their positions as fundamental mechanical units by the metre and kilogram. The unit of time (second) was to remain unchanged. These modifications are contained in the metre-kilogram-second (MKS) system of units.

Giorgi further suggested that some basic relations involving current, charge and field quantities should be modified using the numerical factor 4π. It is this last proposal which has given rise to the most intense debate, and its designation 'rationalisation', although now common usage, cannot be claimed to have endeared it to its opponents.

With the changes suggested by Giorgi, the law of force per unit length between long parallel straight wires *in vacuum* carrying currents i_1 and i_2 amperes is represented as

$$F = \frac{\mu_0 i_1 i_2}{2\pi a}.$$

Here F is the force in newtons and a is the separation in metres. μ_0 is the permeability of free space. It is assigned the value $4\pi \times 10^{-7}$ henry metre^{-1}, and is called the magnetic constant.

It was proposed further that for consistency Coulomb's law should take the form

$$F = \frac{q_1 q_2}{4\pi\epsilon_0 r^2},$$

where F is the force in newtons between point charges *in vacuum* and r is the separation of the charges in metres. The units for q_1 and q_2 are coulombs. It is not difficult to show that with these units the speed of light c in empty space satisfies the relation

$$c^2 = \frac{1}{\mu_0 \epsilon_0},$$

so that

$$\epsilon_0 = \frac{1}{\mu_0 c^2}.$$

This determines the value of the electric constant ϵ_0, which is alternatively called the permittivity of free space.

12.3.4　Advent of MKSA and SI systems

The basis of the MKSA system adopted by the General Congress of Weights and Measures in 1948 was the acceptance of the metre, kilogram and second as the fundamental mechanical units, with the ampere as an additional basic unit. These four suffice for the establishment of units for all the various electromagnetic quantities.

Twelve years later, the same Congress adopted definitions for additional non-electrical quantities, as part of a new and broader International System of Units, or Système International d'Unités (SI).

12.3.5　Magnetic constant. Definition of ampere

The suggestions by Giorgi concerning rationalisation were also incorporated into the MKSA and SI systems. A particular consequence was the acceptance of the value $4\pi \times 10^{-7}$ henry metre^{-1} for the magnetic constant μ_0.

The ampere was formally defined as that constant current which, if maintained in two straight parallel conductors of infinite length, of negligible circular cross-section, and placed one metre apart *in vacuum*, would produce between them a force of 2×10^{-7} newtons per metre of length. Inspection of the formula

$$F = \frac{\mu_0 i_1 i_2}{2\pi a}$$

shows that this definition is consistent with the value assigned to the magnetic constant.

12.4　Establishment of electrical standards

The present-day procedure whereby the necessary representative collection of electrical and associated mechanical standards is established is illustrated in Fig. 12.1. The diagram is in part self explanatory.

The first row in the diagram contains the three basic mechanical units. The combination of quartz oscillator and frequency divider in the second row provides a portable and readily reproducible unit of time derived from the spectrum of caesium-133.

Two quantities which have to be determined experimentally appear in the third row. These are the speed of light c, and the magnitude at appropriate locations of the acceleration due to gravity g. The former is

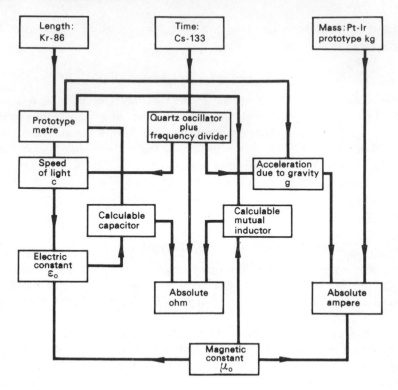

Fig. 12.1 Establishment of electrical standards

required so that a value for the electric constant ϵ_0 can be derived from the magnetic constant μ_0. The latter is needed in any practical realisation of the ampere, because it enables the force between two current-carrying circuits to be measured in terms of the gravitational force acting on a known mass.

In due course we shall see that the calculable capacitor and calculable inductor of the fourth row constitute standards of capacitance and inductance whose values can be derived from knowledge of their physical dimensions, and of the constants ϵ_0 and μ_0 respectively. With the addition of a known frequency or unit of time, either of these can then be used to determine the value of a resistance in absolute measure.

The final step is the production of calibrated standards of capacitance, resistance, inductance and emf, with which sub-standard components can subsequently be compared.

12.5   Realisation of ampere

The formal definition of the ampere obviously cannot be applied directly

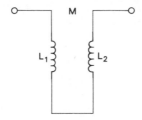

Fig. 12.2 Mutual inductor with series-connected coils

for its practical realisation, in view of the infinite length of conductors that this would entail, quite apart from difficulties associated with the smallness of the force per unit length. It is usual at the present day to work instead with series-connected current-carrying coils of simple and accurately known geometry.

With the usual notation (section 5.3.10), the energy W stored by mutually coupled coils (Fig. 12.2) is given by the relation

$$W = \tfrac{1}{2}L_1 i_1^2 + \tfrac{1}{2}L_2 i_2^2 + M i_1 i_2.$$

If the coils are connected in series, so that the primary and secondary currents are the same and equal to i, say, then

$$W = \tfrac{1}{2}(L_1 + L_2)i^2 + M i^2.$$

In absolute-ampere determinations it is usual for the coils to be coaxial. Let us suppose that one coil is movable along the common axis, and that the coordinate of its position on this axis is x. Let x change by a small amount dx, the current being maintained constant meanwhile. In the absence of ferrous materials, the self inductances L_1 and L_2 remain constant. Then the resulting increase in energy of the coils is

$$dW = i^2 \, dM = i^2 \frac{\partial M}{\partial x} \, dx.$$

The axial force experienced by the movable coil is

$$F = -\frac{\partial W}{\partial x} = -i^2 \frac{\partial M}{\partial x}.$$

This force can be measured with a sensitive balance in terms of the gravitational force exerted on a counterbalancing mass. The current can then be evaluated in absolute measure in terms of the quantity $\partial M/\partial x$, which depends in a calculable way on the geometry of the coils and the magnetic constant μ_0.

A non-absolute ammeter could then be calibrated by passing the current through it and noting the response.

12.5.1   Rayleigh current balance

One method of determining a current in absolute measure uses the Rayleigh current balance. In this instrument (Fig. 12.3) two identical fixed coils are connected as a series-aiding Helmholtz pair, with axial separation equal to their radii. A movable smaller current-carrying coil is suspended from the arm of a balance, coaxial with and midway between the fixed coils.

The procedure consists essentially in determining the change Δm to be made in the size of the counterbalancing mass, which just compensates for a reversal of the current i in the fixed coils. If all three coils are connected in series, then

$$2i^2 \frac{\partial M}{\partial x} = g \, \Delta m.$$

With correct positioning of the coils along the common axis, the quantity $\partial M/\partial x$ actually takes a maximum value, and any necessary slight adjustment of their separation can be performed in the course of a preliminary weighing. The position of symmetry of the movable coil is fortunately then least critical.

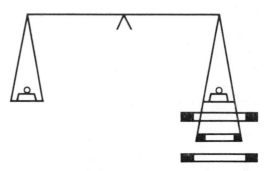

Fig. 12.3 Rayleigh current balance (Redrawn from Duffin, W. J. (1965), *Electricity and Magnetism*, McGraw-Hill Book Company.)

The quantity $\partial M/\partial x$ contains the factor μ_0, but otherwise lacks the dimension of length. For correctly positioned single-turn coils of circular cross-section, $\partial M/\partial x$ depends only on the *ratio* of the radii of the fixed and movable coils. This ratio can be determined in a subsidiary experiment in which the movable coil is coplanar with one of the fixed coils, the two being positioned with their common plane vertical and in the meridian. Differing currents are passed in opposite senses through the two coils, and the ratio of these currents is adjusted so that a magnet suspended at the centre of the coils is undeflected. It is easy to show that the ratio of currents is then equal to the ratio of radii. The former quantity can be measured accurately by a simple potentiometer technique.

The magnetic force produced in the weighing experiment is small unless closely wound coils of many turns are employed. There is then no longer a unique ratio of radii, and it becomes necessary to determine the radius and position of every turn. Despite continuing development, the attraction of the method is therefore greatly diminished.

Notice the very obvious dependence of current-balance determinations on the accuracy with which the acceleration due to gravity is known for the immediate locality of the apparatus.

12.5.2 Ayrton–Jones current balance

The Ayrton–Jones balance differs from the Rayleigh balance in the use of single-layered helical coils (Fig. 12.4), with which axial forces of adequate magnitude are easily produced, and for which the quantity $\partial M/\partial x$ can be readily calculated. Current can be determined to within a few parts per million.

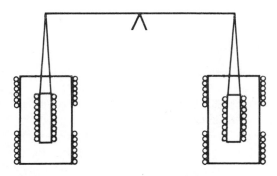

Fig. 12.4 Ayrton–Jones current balance (Redrawn from Duffin, W. J. (1965), *Electricity and Magnetism*, McGraw-Hill Book Company.)

The coils are wound in helical grooves cut on four marble cylinders. The diameter of the bare copper wire is measured repeatedly as winding proceeds. It is essential that the dimensions of the coils be known with great accuracy, and several thousand readings may be needed of the dimensions and positioning of the various turns.

When the suspended coil is positioned symmetrically within the fixed coils, no force is experienced if the currents in the latter circulate in similar senses. An axial force will then develop if the current in one fixed coil is reversed.

The six coils are connected in series in such a way that when current passes, mutually reinforcing couples are produced about the axis of the suspension. If now the connections to the four fixed coils are reversed, the senses of all the couples are reversed, and the necessary change Δm in counterbalancing mass will be given by

$$2i^2 \frac{\partial M}{\partial x} = g \, \Delta m.$$

Here $\partial M/\partial x$ includes contributions by all coils.

Each movable coil experiences an additional force caused by the remote pair of fixed coils. Error due to this effect can be eliminated by reversing all the currents in one trio of coils, which changes the sign of the unwanted force. The two values obtained for Δm are then averaged.

The effects of weak stray fields originating from permanent magnets and electric currents in the vicinity can be eliminated by simultaneously reversing the currents in all six coils and averaging the new and original Δm values.

The influence of materials with appreciable magnetic susceptibility cannot be eliminated in this way. The balance and its heavy pedestal are therefore made of bronze, and any material which might affect the forces between the coils is removed from the vicinity of the apparatus.

12.6   Realisation of ohm

The electrical resistance of a sample of material is determined by its dimensions and the shape and positions of the electrodes, and by the resistivity of its material. This last quantity is a physical property, and cannot be predicted from knowledge of its non-electrical properties.

By contrast, the reactance at a chosen frequency of a suitably designed inductor or capacitor can be calculated with great accuracy from a knowledge of only the geometry, and of the conventionally assigned properties of the surrounding medium, provided this is vacuum. Absolute-ohm determinations therefore take the form of a comparison of a resistance with such a known reactance.

12.6.1   Lorenz disc

The method described below was proposed by Lorenz in 1873. The apparatus consists basically of a circular metal disc located inside and coaxially with a uniform solenoid (Fig. 12.5). The disc is rotated at high speed about its axis by an electric motor. The solenoid is connected in a closed series circuit with a battery, a rheostat, and the resistance R which is to be determined in absolute measure. Electrodes are provided in the

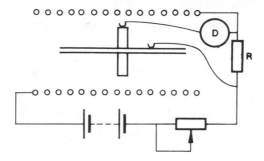

Fig. 12.5 Basic Lorenz disc apparatus

form of brushes which make contact with the rim of the disc and the spindle connecting it to the motor.

The mutual inductance between two circuits is the emf induced in one per unit rate of change of current in the other. In the Lorenz disc apparatus, the current i in the solenoid is steady, but an emf develops between the rim and axis of the disc on account of its rotation. The magnitude of this emf can be shown to be Min, where n is the frequency of rotation of the disc and M is the mutual inductance between solenoid and disc. This latter quantity is proportional to the magnetic constant μ_0 and a geometrical factor determined by the dimensions of the solenoid and disc. The Lorenz method therefore depends ultimately on a calculable mutual inductance.

The emf generated in the disc is balanced against the potential difference iR developed across the resistor, the condition for equality being

$$R = Mn.$$

The magnitude of the induced emf is small, so that sources of error have considerable effect and must be carefully minimised.

In the apparatus due to F. E. Smith (1913), two discs are mounted in separate solenoids on the same spindle (Fig. 12.6). The solenoids are wound with bare copper wire laid in helical grooves cut in marble cylinders, and both are split in the same way as the outer coils in the Ayrton—Jones current balance. The emf is extracted between the rims of the two discs. The current is circulated in opposite senses in the two solenoids, so that the induced emfs are additive, whereas cancellation occurs for thermo-emfs and contact potentials. The planes of the discs are aligned with the Earth's magnetic field, and any small induced emfs caused by misalignment should cancel mutually. The electric motor is positioned some distance away, so that magnetic interaction with the apparatus is minimised. The rim of each disc is in zero magnetic field, and the radii are therefore not critical. In an ideal situation the speed of rotation would be

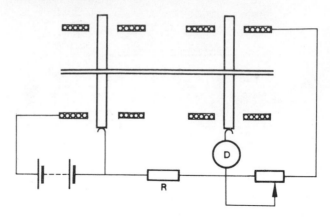

Fig. 12.6 F. E. Smith's version of Lorenz disc apparatus

adjusted for zero detector current. In practice the effects of stray emfs cannot be entirely eliminated, and the speed chosen is that for which the galvanometer deflection is unchanged by reversal of the current in the solenoids. Contact resistance at either disc is obviously of little consequence, as only a small current is drawn. Stability of the frequency of rotation n is essential to ensure that it can be measured accurately by stroboscope or chronograph.

The Lorenz disc provides an interesting technique, but is at present of no more than historical significance. The probable uncertainty provided by the method is nevertheless only a few parts per million, which is similar to the uncertainty involved in current-balance measurements.

12.6.2   Campbell calculable mutual inductor

The standard mutual inductor developed at the National Physical Laboratory by A. Campbell is constructed with a split single-layer primary, the two halves being wound on a fused quartz cylinder and separated by a distance approximately equal to twice the length of each (Fig. 12.7). The secondary consists of several hundred turns, wound on a former of larger diameter than that of the primary and positioned centrally and coaxially with it. The arrangement gives a mutual inductance of several millihenries. The actual value can be calculated to about one part per million, and varies slowly and predictably with change in diameter of the multi-layer secondary.

Realisation of the ohm is carried out using the circuit of Fig. 12.8. The continuously variable mutual inductor M_2 has been previously calibrated against the calculable mutual inductor M_1. When the detector

Fig. 12.7 Campbell primary standard of mutual inductance (Redrawn from Vigoureux, P. (1971), *Units and Standards for Electromagnetism,* Wykeham Publications.)

indicates a null response, Kirchhoff's second law gives for the detector loop the relation

$$j\omega M_2 i + r i_0 = Si.$$

Let R and L be the total series resistance and inductance respectively in the closed circuit linking the two mutual inductors. Then Kirchhoff's second law gives for this

$$Ri + j\omega Li + j\omega M_1 i_0 = 0.$$

Elimination of the currents i and i_0 between these relations gives easily

$$Lr = -M_1 S,$$

and

$$rR = -\omega^2 M_1 M_2.$$

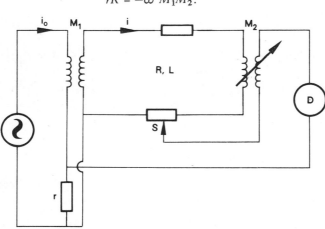

Fig. 12.8 Campbell bridge network for realisation of absolute ohm (Redrawn from Vigoureux, P. (1971), *Units and Standards for Electromagnetism,* Wykeham Publications.)

Balance is obtained by adjustment of M_2 and S. The second balance relation gives a value for the product Rr, and the resistance to be standardised is determined by comparing it first with R and then with r. The probable error in the standardised value is only about two parts per million.

12.6.3   Calculable capacitor

Until relatively recent times it appeared to be impossible to design a calculable capacitor which would provide the high accuracy required in absolute measurements. Standard capacitors had therefore to be calibrated against a known resistance in terms of a known frequency.

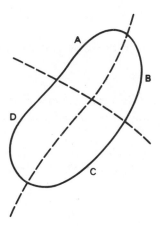

Fig. 12.9   Cross-section of segmented cylinder

This situation was reversed in 1956 following the work of Thompson and Lampard, who pointed out that for any infinitely long right conducting cylinder *in vacuum*, which is divided into four segments A, B, C and D as in Fig. 12.9, the capacitances per unit length C_1 and C_2 between alternate segments A, C and B, D respectively are connected by the relation

$$\exp \left(-\frac{\pi C_1}{\epsilon_0} \right) + \exp \left(-\frac{\pi C_2}{\epsilon_0} \right) = 1.$$

If now these cross-capacitances C_1 and C_2 are arranged to be equal, the value of either is given by the simple formula

$$\frac{\epsilon_0}{\pi} \log_e 2.$$

The importance of this result is that it is independent of the shape of the

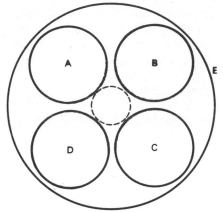

Fig. 12.10 **Practical calculable capacitor** (Redrawn from Vigoureux, P. (1971), *Units and Standards for Electromagnetism*, Wykeham Publications.)

cross-section and the locations of the segment boundaries. Even if C_1 and C_2 differ slightly in magnitude, the error in the formula is small and can be computed with acceptable accuracy.

A version of the calculable capacitor suggested by Thompson and Lampard has been constructed at the National Physical Laboratory. It consists basically of four similar parallel metal rods A, B, C and D, mounted in close mutual proximity (Fig. 12.10). These are shielded from nearby apparatus by the cylindrical, earthed, metal tube E. The symmetry of the arrangements ensures near-equality of the cross-capacitances.

A pair of earthed metal tubes is inserted centrally, one from each end of the system, at the position occupied by the broken circle and in close proximity to the four rods. The screening action of these tubes eliminates cross-capacitances between the rods along the lengths flanked by the tubes.

Errors due to electrical end-effects are eliminated by varying the separation between the inner ends of the central tubes. Changes in this separation are determined using a photoelectric technique for counting interference fringes formed by multiple reflection of laser light travelling axially within the interspace. This light enters and leaves the system via the hollow central tubes.

Cross-capacitance provided by the calculable capacitor amounts to only a fraction of a picofarad. This is inconveniently small, and a simple transformer ratio-arm bridge is used in successive capacitance steps of $10:1$ to compare it ultimately with two stable capacitors providing nominally about 5 nF each.

The transformer bridge of Fig. 12.11 is used to relate capacitance and resistance in terms of frequency. The components C_0 and C_0' are the

5 nF capacitors. The stable resistors R_0 and R_0' are as near identical as possible, and their nominal resistance values of 20 kΩ approximate to the reactance of a 5 nF capacitor at an angular operating frequency of 10^4 rad s^{-1}. The quantities v_1, v_2 and v_3 are the complex voltages established between the locations indicated and the earthed centre-tap of the transformer. When both detectors indicate null readings, junctions E_1 and E_2 are at earth potential, so that

$$i_2 = \frac{v_1}{R_0} = -j\omega C_0 v_2,$$

and

$$i_3 = j\omega C_0' v_3 = -\frac{v_2}{R_0'}.$$

Rearrangement gives

$$v_2 = \frac{j v_1}{\omega C_0 R_0} = -j\omega C_0' R_0' v_3,$$

and

$$\frac{v_1}{v_3} = -\omega^2 C_0 C_0' R_0 R_0'.$$

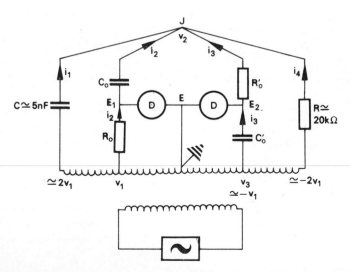

Fig. 12.11 Transformer bridge for comparison of capacitance and resistance in terms of frequency (Redrawn from Vigoureux, P. (1971), *Units and Standards for Electromagnetism*, Wykeham Publications.)

These relations indicate that v_1, v_2 and v_3 should be approximately equal in magnitude, that v_1 and v_3 should be in antiphase, and that v_2 should lead v_1 by $\pi/2$ radians. The use of a transformer ensures the desired phase relationship between v_1 and v_3, while that for v_1 and v_2 is arranged by adjustment of the subsidiary components C and R. The feasibility of the latter situation can be verified as follows. The total current arriving at J is necessarily zero, so that

$$i_1 + i_2 + i_3 + i_4 = (2v_1 - v_2)j\omega C - j\omega C_0 v_2 - \frac{v_2}{R_0'} - \frac{(2v_1 + v_2)}{R} = 0.$$

In view of the near-equality of $1/\omega C$, $1/\omega C_0$, R_0' and R, this reduces to

$$\frac{v_1 + v_2}{v_1 - v_2} = j,$$

approximately, or

$$v_2 = jv_1,$$

approximately, as required.

The angular frequency chosen for the source is 10^4 rad s^{-1}. The bridge is balanced by adjustment of C and R, and by alteration of v_3 by known amounts with the aid of an auxiliary transformer. It is essential that the ratio v_1/v_3 be known accurately, unlike the values of C and R, which do not directly affect the standardisation.

The resistors R_0 and R_0' are subsequently connected in parallel, and a transformer bridge is used to compare the combination with a 10 kΩ resistor for which the dependence of resistance on frequency is known. This is compared in turn with a 1 Ω laboratory standard resistor, using a special Wheatstone bridge. Despite the several steps involved in relating the calculable capacitor to this resistor, the overall uncertainty involved is reckoned to be less than one part per million.

12.7   Material standards

Absolute measurements like those just described require elaborate equipment, and take up a great deal of the time of highly skilled persons. Although improvement and development of apparatus and techniques are more or less continuous, updating of the value in absolute units of a given primary standard is therefore not usually undertaken more than once in a period of several years. The values of the more important electrical quantities are maintained meanwhile in terms of calibrated material standards.

It is not feasible to use an instrument with a mechanical or optical pointer for the preservation of an electrical unit, the inherent uncertainties in such devices being unacceptably large.

12.7.1   Standard resistors

The ohm is conveniently preserved directly, using standardised resistors. The highest quality standards are constructed with bare annealed wire of an alloy selected for its low temperature coefficient of resistance and low thermal emf against copper (see also section 4.2.1). The wire must be self-supporting, and is therefore of thick gauge, so that the resistance is likely to be no greater than one ohm. It is hermetically sealed in oil at a thermo-statically controlled temperature, and provided with separate current and potential leads to avoid errors caused by variation in contact resistance at the terminals.

For international comparisons, stability of resistance approaching one part in 10^7 is considered to be essential. Resistance values tend to drift in the first few years of life, and instability reappears with advancing age. It is customary for standards laboratories to maintain a number of similar resistors as a reference group, the average resistance of a batch being more reliable than that of any individual resistor.

12.7.2   Standard cells

The Weston cadmium cell (section 4.1) provides a stable standard of emf when constructed with materials of high purity. Saturated cells are favoured for stability with time, although unsaturated versions exhibit a lower temperature coefficient of emf. Constancy of temperature of the saturated cell is essential, even when the cell is not in use, to ensure maximum stability at all times. Like standard resistors, standard cells are maintained in batches, and greater weight is attached to the average emf than to individual values. EMFs tend to vary when the cell is relatively young, and the behaviour eventually deteriorates with age. The useful life is about thirty years.

Uncertainty in the emf provided by the Weston cell amounts at best to several parts per million. The cell is of course restricted to dc measurements, whereas standard resistors can be used with alternating currents of low frequency.

12.7.3   Zener diode

The Zener diode is a silicon semiconductor device. Over a suitable range of reverse conduction currents, the potential difference which develops is extremely stable. If provision is made for stabilisation of the current, the device can be used as a standard of emf which is not greatly inferior in constancy to the Weston cell. Operating voltages are of the order of several volts, and the current passed by the diode is a few milliamperes.

12.7.4   Josephson effect

If two superconductors are separated by a thin film of insulating material, such as a metallic oxide, conduction can occur through the gap as a result of what is called *tunnel effect.* The conduction process does not normally give rise to any potential difference between the superconductors. If however the junction is irradiated with electromagnetic waves of angular frequency ω, a potential difference develops which increases in uniform steps as the current increases. The height V of each step of potential is related to ω by the equation

$$4\pi eV = h\omega,$$

where e is the magnitude of the electronic charge and h is Planck's constant.

The electromagnetic radiation can be derived from a klystron, or a long-wave laser. The effect seems capable of providing an extremely stable comparison voltage, but the technique is not easy because V is quite small in presently realisable situations.

12.7.5   Gyromagnetic ratio of proton

In a magnetic field B, the axis of spin of a particle having magnetic moment m is either parallel or anti-parallel to the direction of B, corresponding particle energies being $-mB$ and $+mB$. In a transition from one state to the other, the energy change is $2mB$. Now energy is emitted or absorbed in quanta, so that the frequency ν of the associated radiation is given by

$$h\nu = 2mB.$$

If ω is the corresponding angular frequency, this can be written alternatively in the form

$$\omega = \frac{4\pi m}{h} B.$$

The factor $4\pi m/h$ is called the *gyromagnetic ratio* γ, and is a constant for a given type of particle in a given material.

The effect can be used for the preservation of the ampere, the particles being protons in a volume of water. In the stable state the spin axes are parallel to the magnetic field, and if the direction of the field is changed a relaxation time is involved. If the field is provided by a solenoid, its magnitude B is calculably related to the steady current i which is flowing. The relationship between B and the angular frequency of the associated radiant energy therefore gives means of checking the constancy of i. It is even conceivable that the method may rival the current balance as a means of realisation of the ampere.

There are two ways of utilising the effect. The first is a weak field technique. A strong magnetic field is provided initially, and this produces a relatively large magnetic bias. The strong field is now switched off, leaving a weak calculable field provided by a uniform solenoid, which is applied at right-angles to the direction of the strong field. For two or three seconds there is a decay of bias caused by intermolecular collisions in the water, with consequent radiation of electromagnetic energy of angular frequency related to the intensity of the weak field by the above relation. With suitable choice of B, the frequency is of the order of 50 kHz, and can be fairly easily measured.

In the alternative strong field technique, an alternating VHF field is provided by a subsidiary coil. This stimulates the transitions, and an absorption resonance occurs if the frequency is suitably related to the intensity of the strong steady magnetic field provided by a large magnet. In practice it is best to modulate this field by winding on the magnet a few additional turns of wire carrying current of mains frequency. A quantity related to the absorption of the VHF field is displayed as the vertical deflection of the trace on a cathode-ray oscilloscope, the horizontal deflection being proportional to the instantaneous value of the 50 Hz modulating current. With the intense field provided by the magnet, a VHF frequency of the order of 50 MHz is required.

12.7.6   Standard capacitors

Good-quality capacitors have the advantage of virtually negligible dissipation of energy, in contrast with inductance, and more especially with resistance, so that problems of inherent temperature-rise under conditions of use are correspondingly diminished. Standard capacitors are constructed using either silvered mica, or interleaved mica and metal foil. Hermetic sealing is essential to prevent ingress of moisture, and the temperature is thermostatically controlled to avoid associated changes in linear dimensions.

A calculable capacitor has already been described (section 12.6.3).

This is of course a permanent standard, and can be used in conjunction with a determined frequency to preserve the ohm.

12.7.7 Standard inductors

Both self and mutual inductors are maintained as material standards. Ferrous cores are impractical because of nonlinearity and the marked hysteresis effect, and the coils generally are air-cored, and are wound on a marble or ceramic former. For high-frequency operation the turns and layers are widely spaced to minimise self-capacitance. As the frequency rises, the current is increasingly restricted to the surface layer of the conducting wire. This is skin effect, which causes the effective inductance to diminish while the series resistance increases. Skin effect precludes the use of metal screening cans, unless the field of the coil is strictly localised by toroidal winding.

12.7.8 Calibration

In major developed countries, some standards laboratories are maintained by the State and others by the larger industrial organisations. Each is equipped to provide a calibration service against its own material standards.

In the United Kingdom, the *house standards* of each industrial standards laboratory are checked periodically against those of the National Physical Laboratory. International comparisons of electrical standards are carried out at intervals of a few years. The results are published, but the units of the participants are adjusted much more rarely, because in the relatively short term continuity is valued more than precision.

The Wheatstone bridge can be used for the comparison of resistances ranging down to one ohm, while the Kelvin double bridge is better suited for still smaller values. The resistive potential divider can also be used for this kind of comparison, as well as for voltage calibrations against the emf of the Weston cell. Alternating current is used increasingly for component calibrations, both for convenience and because of the growing need for calibration over a wide range of frequencies. The resistive potential divider is being displaced increasingly by the inductive divider. The latter was first suggested as an alternative to the resistive ratio arms of the Wheatstone-type bridge towards the end of the last century. With careful design, the voltage ratio is closely equal to the turns ratio, and extremely high accuracies are therefore available for the comparison of component values in its development as the transformer bridge.

The establishment of an ac calibration, proceeding from the starting point of the corresponding dc calibration, is relatively easy for resistance so long as the frequency is low, but there are obvious difficulties for voltage and current. Electrostatic voltmeters can be employed for the transference of voltage calibrations from dc to ac. Both current and voltage transference can be effected with the aid of thermojunction instruments, in which the heating effect caused by the flow of dc and ac currents is arranged to generate equal thermal emfs.

It is noticeable that independent methods are available for maintaining the volt, the ampere, and the ohm, and as any two of these suffice for the determination of the third, a check can be made on the consistency of the three.

Additional reading

BALDWIN, C. T., *Fundamentals of Electrical Measurements*. Harrap, 1961.

DUFFIN, W. J., *Electricity and Magnetism*. McGraw-Hill, 1965.

GOLDING, E. W. and WIDDIS, F. C., *Electrical Measurements and Measuring Instruments*. Pitman, 1963.

HARNWELL, G. P., *Principles of Electricity and Electromagnetism*. McGraw-Hill, 1949.

HARRIS, F. K., *Electrical Measurements*. Chapman and Hall, 1952.

THOMPSON, J. R., *Precision Electrical Measurements in Industry*. Butterworths, 1965.

VIGOUREUX, P., *Electrical Units and Standards*. HMSO, 1970. (National Physical Laboratory.)

VIGOUREUX, P., *Units and Standards of Electromagnetism*. Wykeham Publications, 1971.

AMERICAN INSTITUTE OF PHYSICS HANDBOOK, 2nd edition, McGraw-Hill, 1963, 5—98.

SPECIAL PUBLICATION, NBS No. 300, Vol. 3. HERMACH, F. L. and DZIUBA, R. F. (eds.) *Precision Measurement and Calibration. Electricity — Low Frequency*. 1969.

Answers

Chapter 2:
1. 50 mA.
2. 360 C.
3. 21 min.
4. 63 p.
5. 0.2 A, 200 V.
6. 0.25 A, 800 Ω, 180 kJ, 900 C.
7. 250 A, 500 A.

Chapter 3:
1. 2 Ω.
2. 20 mS.
3. 10^{-4} m².
4. 3 Ω, 1 A, 2 A.
5. 100 Ω.
7. 13 Ω.
8. 1 A, 1 A, 1 A, 1 A.
9. 4 Ω, 2 V.
10. 24 V.
11. 0.1 A.
12. 70 W.
13. 0.5 A, 2 A, 0.5 A, 0.25 A.
14. (*a*) 6 A, 2 A, 4 A. (*b*) 1 A, $\frac{1}{3}$A, $\frac{2}{3}$ A. (*c*) 0.25 A, 0.25 A, 0.5 A. (*d*) 2 A, 1.75 A, 0.25 A. (*e*) 2.5 A, 2 A, 0.5 A. (*f*) 1 A, 1 A, 0 A.
15. 1 A, 1 A, 0 A.
16. 3 Ω, $\frac{1}{2}$; 4 Ω, $\frac{1}{3}$; 5 Ω, $\frac{2}{3}$.
17. No.
18. 8 Ω, $\frac{1}{3}$, 20 $\log_{10}$ 3; 36 Ω, $\frac{1}{5}$, 20 $\log_{10}$ 5; 3, $\frac{1}{4}$, 40 $\log_{10}$ 2; 12 Ω, $\frac{1}{5}$, 20 $\log_{10}$ 5; 4 Ω, $\frac{1}{9}$, 40 $\log_{10}$ 3.

19.   Series resistors 20 Ω, shunt resistor 30 Ω.

Chapter 4:

1.   10 mA.
2. $v_0/(1 + R/r)$ to v_0, 20 mA.
3.   20 mA, 50 Ω and 200 Ω from one end, 15 mA and 30 mA.
4.   0.32 A, 1.28 W.
5.   0.2 V.
6.   0.1 MΩ, 1 mA.
7.   1 mA, add 9 kΩ.
8.   Provide 10 Ω, 190 Ω, 1990 Ω in series.
9.   3 Ω shunt, 2 Ω.
10.   10 Ω, 1 Ω and 0.5 Ω shunts.
11.   Shunt series-connected resistors are 5, 9, 0.9, 0.09 and 0.01 Ω.
12.   82 Ω.

Chapter 5:

1.   180 000 F.
2.   1 kV, 1 μJ, 10 μJ.
3.   7 μF.
4.   4 V, 48 μJ.
5.   1 μF.
6.   36 μC, 18 μC, 6 V each.
7.   2 ms, 40 ms, 100 μA.
8. e^{-2}.
9. $C(R_1 + R_2)$, CR_2.
10.   3 s, 2 s.
11.   6 s, 4 s.
12. $8(1 - e^{-1})$ μC, $4(1 - e^{-1})(1 + 2e^{-1})$ μC.
13.   200 μV.
14.   100 J.
15.   110 mA, 1.1 A, 1100 V, opposite, 27.5 J.
16.   0, 200 As^{-1}.
17.   0, 1500 As^{-1}, 10 A.
18.   400 ms, 300 ms.
19. $v_0(2 - e^{-1})/R$.
20.   20 mV.
21.   1.2 J.
22.   80 V.
23.   5 Ω, 2.5 V.
24.   0 A, 500 As^{-1}, 2.5 V.
25.   19 Ω.

26. $10 \, \Omega$.
27. e^{-1} A, 100 μs.
28. $0.05 \, e^{-1}$ A.
29. 50 divisions.

Chapter 6:

1. $(\sqrt{(2)}/5\pi)$mC, $(10\sqrt{(2)}/\pi)$C.
2. 2 A, $2\sqrt{2}$ A, $4\sqrt{2}$ A.
3. $7.5 \, \Omega$, 20 A.
4. 120 W, 0.5 A.
5. $8/\pi$ A.
6. 200 W, 1 A, $\sqrt{2}$ A, 0 A.
7. 120 V, 30 V.
8. 10 mA, 0.4 W.
9. 80 mA, 0.48 W.
10. 180 mA.
11. 0.3 mS, 3 mA.
12. 1 A.
13. 40 p, 20 p, 5 A.
14. 0.9 A, 100 μA, 30 mW.
15. 10^4 rad s^{-1}, 1 V, 1 V, 1 V.
16. $(1-j)$ mS, $\sqrt{2}$ mS, 1 mS, -1 mS; $(1+j)$ mS, $\sqrt{2}$ mS, 1 mS, $+1$ mS.
17. 1/338 H, 10 kΩ.
18. 1 H, 4 μF.
19. (*a*) 0.25 MΩ, 100 μA, (*b*) 13 ω_0/12.
20. 100/96 μF.
21. 500 Ω, 2 μF, 0.5 mW.
22. $\frac{1}{2}(1 + e^{-\pi})$ A.
23. 10^4 rad s^{-1}.
24. 35 mH *or* 85 mH.

Chapter 7:

1. 20 μF.
2. 12.5 mH.
3. 2 μF, 2500 $\sqrt{2}$ rad s^{-1} upwards.
4. 25 mH, 2000 $\sqrt{2}$ rad s^{-1}.
5. $0 \, \Omega$.
6. $20\sqrt{50} \, \Omega$.
7. 1 mA.
8. 0.5 W.
9. $75 \, \Omega$.

    10.   4.

    11. $L = 0.2\ \mu\text{H m}^{-1}$, $C = 1/18\ \text{nF m}^{-1}$.

    12. Z_0^2/Z_L, 120 Ω.

Chapter 8:

    1.   10 kΩ.

    2.   200 μH, 1.

    3.   200 rad s^{-1}, 50 nW.

    4.   1 nF, 2 Ω.

    5.   40 μH, 2 Ω.

    6.   100 μH, 1 Ω, 1 MΩ.

    7.   2 mH, 100 mV, 50 μA.

    8. 5×10^6 rad s^{-1}, 500 μA, 50, 25 kΩ.

    9.   50 μH, 500 pF, 10 Ω, 0.5 μA, 5 π μA.

Chapter 9:

    1.   5 mH.

    2.   100 μA, 300 μA.

    3. $\pi/4$.

    4.   50 Ω.

    5.   100 nF.

    6.   0.5 μA, 15 pW; 4860 Ω, 0.2 mH.

Chapter 10:

    1.   0.1 H, 30 Ω, 500 Ω, 0.4 μF, 500 Ω, 4 μF.

    2.   0.1 μF, 1 Ω, 10 000 rad s^{-1}.

    3.   100 Ω, 50 pF, 125 pF, 62.5 Ω, 0.078.

    4.   0.4 μH, 1 Ω.

    5. 10^8 rad s^{-1}.

Chapter 11:

    1.   10 μA.

    2.   392 Ω.

    3.   0.25%.

    4.   0.25 A, 0.2 A.

Index

absolute balance, 213, 223
acceleration due to gravity, 263−4, 267
admittance, definition of, 144
aggregate balance, 213, 223
ammeter, 48−51, 59, 106, 128−30
Ampère, André Marie, forces between currents, 5
ampere,
  definition of, 8, 20−1, 261, 263−4
  preservation/realisation of, 43, 264−8, 227−8, 280
amplitude factor, 127
Anderson L/C bridge, 222−3
angular frequency, 124
anisotropy, electrical, 26
aperiodic transient response, 103
apparent wattage, 134
Arago's disc, 6
Argand diagram, 141−2
argument of complex number, 138−9, 141
attenuator, 36−8
autotransformer, 194, 196
Ayrton−Jones current balance, 267−8, 269
Ayrton universal shunt, 50−1

back-emf, 86
balanced and screened transformer, 225, 229−30
balanced transmission line, 163
ballistic bridge, 213, 221, 222, 223
ballistic galvanometer, 70, 84, 106−7, 113−15, 213
bel, 37
branch current, 28
bridge detector, 240−1
bridged-T bridge, 233−9, 255−6
bridge principle, 180, 213, 232
bridge source, 240−1

Campbell, A., mutual inductor, 270−1
capacitance,
  measurement of, 218−19
  unit of, 3, 70
capacitor,
  behaviour of,
    for ac, 130−2, 137
    for dc, 69−70
  calculable, 264, 268, 272−5, 278−9
  charging through resistance, 69−70, 77−9, 80−2
  commercial, 82−3
  complex impedance of, 137, 139
  continuously variable, 82
  dielectric in, 70, 82−3, 278
  discharging through resistance, 75−7, 80, 83
  electrolytic, 83, 145
  energy storage by, 70−2, 82, 88, 99, 104, 115, 132, 184
  geometry, influence of, 70
  impedance of, 132, 139
  imperfect, 82, 144−6, 218−19
  phase angle for, 131, 139
  power factor for, 134
  standard, 82, 97, 264, 278
capacitors,
  parallel combination of, 72−3
  series combination of, 73−5
carbon arc, 4, 7
carbon filament lamp, 7
Carey−Foster M/C bridge, 223−4
Carey−Foster resistance bridge, 62−3
CGS electrical units, 8, 259−62
characteristic impedance, 164−5, 167, 168−9, 172
charge,
  conservation of, 13, 127, 243
  unit of, 3, 260, 262

286   *Index*

charges,
  comparison of, 113—14
  forces between, 3, 259—60, 262
Christie, Samuel, originator of Wheatstone bridge, 59
circuital current, 28
Clark cell, 4
coaxial line, 163, 168—9
compass, magnetic, 2, 5
Compensation Theorem, 250—1
complex admittance, 143—4
complex conjugate, 147
complex exponent, 135—6
complex impedance, 137—9, 141, 143
conductance, 24, 26, 144
  measurement of, 182—3, 239
conductances, parallel combination of, 24, 128
conductivity, 25—6
conjugate branches, 249
conjugate match, 147
conservation of charge, 13, 127, 243
conservation of energy, 19, 95—6, 127, 243
Coulomb, Charles, inverse square law, 3, 259—60, 262
coupling factor, 195—6, 209—11, 218
crest factor, 127
critical damping,
  of LCR series combination, 106—13
  of moving-coil instrument, 50—1, 106
Crompton potentiometer, 54—5
current,
  scale of, 10—11, 59
  unit of, 8, 20—1, 260—8, 280
current balance, 5, 266—8
current gain, 185, 186—7, 188, 191
current resonance, 175—6
currents, forces between, 5, 20—1, 129, 260, 262—8

damping resistance, 50, 61, 114
Daniell cell, 4
d'Arsonval instrument, 5—6, 129—30
Davy, Sir Humphry, 4
deadbeat transient response, 103
decibel, 37
detector, for ac bridge, 240—1
dielectric breakdown, 82, 93, 172
dielectric loss, 82, 145, 163, 164, 185, 218—19
differentiating network, 115—17
dispersion in transmission line, 168
dominant bridge component, 217
dust-cored coil, 196
dynamo effect, 6—7, 85, 268—70
dynamometer-type instrument, 5, 129

earthing in Wheatstone-type bridge, 213—14
eddy current, 6, 97, 106, 196, 225
Edison, Thomas, filament lamp, 7
electric arc, 4, 7, 93
electric constant, 262—3, 264, 272
electric light, 7
electrochemical equivalent, 12
electrodynamometer, 5, 129
electrolysis, laws of, 4, 12—13
electrolytic capacitor, 83, 145
electromagnetic induction, 6—7, 83—5, 268
electromotive force (emf), 16—17
  thermal, 45, 54, 61, 63, 280
  transformer-induced, 83—5
electrostatic voltmeter, 280
energy, conservation of, 19, 95—6, 127, 243
energy, electrical, 18, 126
equal match, 147

farad, the, 70
Faraday, Michael,
  disc apparatus, 7
  dynamo effect, 7
  laws of electrolysis, 4, 12
  transformer effect, 7, 83, 85
ferrous core, 85, 97, 129, 194, 195, 196, 240, 279
filament lamp, 7
filter, 157—62
former, conducting, 50, 106
  insulating, 106—7
form factor, 127
Fourier component, 122, 167
Franklin, Benjamin, 2—3
frequency,
  definition of, 122, 124
  measurement of, 219—21, 237—8
friction machine, 1, 2, 3, 4, 82
fundamental frequency, 122

Galvani, Luigi, frogs' legs, 3
Giorgi, rationalisation and MKS system, 8, 262—3
Gray, Stephen, 1—2
Guericke, Otto von, friction machine, 1
gyromagnetic ratio, 277—8

half-cycle average, 127
half-power points, 177—9, 187—9, 191
'head effect', 227
Helmholtz pair, 266
henry, the, 85, 86
hertz, the, 124

hot-wire instrument, 128—9
house standard, 279

identical match, 147
impedance, 138—9
  of capacitor, 132, 139
  of inductor, 133, 139
imperfect capacitor, 144—6, 218—19
incandescent lamp, 7
inductance, *see* self inductance, mutual
    inductance
inductor, *see* self inductor, mutual in-
    ductor
instruments, for ac, 128—30
integrating network, 115—17
International Units, 261
iron-cored coil, 85, 97, 129, 194, 195,
    196, 240, 279
isolating transformer, 194, 224

Josephson effect, 277
joule, the, 20, 259

Kelvin, Lord, *also* Sir William Thomson,
    8
  current balance, 5
  double bridge, 63—4, 279
Kelvin—Varley potential divider, 55—6
kilogram, the, 19, 259, 264
Kirchhoff's first law, 12, 13, 28, 127,
    243
Kirchhoff's second law, 15, 17—18, 19,
    28—9, 127, 243

ladder network,
  reactive, 157—62
  resistive, 34—8, 157, 167
laminated core, 97
Lampard, D. G., calculable capacitor,
    272—3
leakage inductance, 200, 232
Lecher line, 163, 169
Leclanché cell, 4
Lenz's law, 86
Leyden jar, 2—4, 82
linear circuit component, 134, 135, 243,
    245
linear current, 28
local action, 4
lodestone, 2
logarithmic decrement, 106
loop current, 28
Lorenz disc, 268—70

magnetic constant, 262—6, 269
magnetising current, 199

matching load, 34—5, 157—8, 159, 162,
    165, 168—9, 172
matching transformer, 195, 229
maximum power theorem, 32—3, 62,
    146—7, 195, 229, 243
Maxwell, James Clerk,
  L/C bridge, 222
  L/L₀ bridge, 221—2
  ratio of units, 260
mechanical units, 19—20, 258—9
mercury ohm, 261
mesh current, 28
metre, the, 19, 258—9, 264
MKSA system of units, 258, 263
MKS system of units, 8, 262
modulus of complex number, 141
moving-iron instrument, 129
multi-test instrument, 51
mutual inductance, 83—5 (*see also* trans-
    former)
  continuously-variable, 96
  measurement of, 217—18, 223—4
  sign convention for, 93—4, 197
  unit of, 85, 96
mutual inductor,
  calculable, 264, 268—70
  energy storage by, 95, 265
  standard, 264, 279

neper, 37
net current, 28
newton, the, 20, 259
non-ohmic material, 26
Norton's Theorem, 248—9

Oersted, Hans, forces between currents,
    5
ohm,
  definition of, 8, 21, 261
  realisation of, 268—72, 275, 279, 280
Ohm's law, 15—16, 26
Owen bridge, 215—18, 230

parallel resonance, 151, 185—92, 201—2,
    232
pass band,
  of reactive ladder network, 157,
    161—2, 174
  of tuned transformer, 210—11
peak factor, 127
peak voltage, 127, 129—30
Pearson, George, electrolysis of water,
    3—4
period, 123
permittivity, 263

phase angle,
  for pure capacitor, 131, 139
  for pure inductor, 133, 139
phase constant, 124
Poggendorff potentiometer, 5
polarisation, chemical, 4
potential, absolute electric, 3
potential difference,
  definition of, 15—16, 17
  unit of, 8, 13—14, 20, 21, 261, 280
potential leads, 45, 276
potentiometer, 51—9
  Crompton, 54—5
  measurements with, 58—9
  modern, 55—8
  Poggendorff, 5
  simple, 52—4, 246—7
Pouillet, tangent galvanometer, 5
power,
  ac, 125—6, 134
  dc, 18
power factor, 134
  of imperfect capacitor, 145—6, 219
  measurement of, 219
power transfer theorem, 32—3, 62,
  146—7, 195, 229, 243
practical units, 8, 261
pulsatance, 124

Q-meter, 180—3
quality factor, Q, 177, 179, 183—4, 188,
  191—2, 232

rationalisation, 262—3
ratio of units, 260
Rayleigh current balance, 266—7
reactance, definition of, 141
Reciprocity Theorem, 249—50
rectifier instrument, 51, 129—30, 240
reflected impedance, 198—9
resistance, 15—16
  definition of,
    for ac, 128, 141
    for dc, 16
  power in, 18, 125—6, 134
  scale of, 16
  specific, 25, 268
  temperature coefficient of, 45, 49—50
  unit of, 8, 16, 21, 261, 264, 268—72,
    275, 280
resistances,
  parallel combination of, 23—4, 127—8
  series combination of, 24—5, 127—8
resistivity, 25—6, 268
resistor,
  bobbin, 45

card, 45, 46
composition, 47, 48
decade box, 46
film, 47, 48
four-terminal, 45, 58—9, 63—4, 276
precision, 45, 46, 47
radio-type, 46—7
standard, 43, 44—5, 264, 276
variable, 47—8
wire-wound, 45, 47, 48
resonance, 174—92, 201—2
  of parallel impure LC combination,
    189—92
  of parallel LC combination, 184—5
  of parallel LCR combination, 185—9
  of series LCR combination, 174—80
rheostat, 48
Robinson frequency bridge, 219—21
root-mean-square (rms) value, 126—7

Schering bridge, 218—19, 230
Schweigger multiplier, 5
screened transformer, 225, 228—30, 232
screening,
  electrostatic, 83, 162—3, 214, 219,
    224—31, 232, 235, 273
  magnetic, 162, 225
second, the, 19—20, 259, 263—4
selectivity, 174, 179, 191, 201
self inductance,
  decade box, 96—7
  measurement of, 215—17, 221—3,
    235—7, 238—9, 255—6
  sign convention for, 86
  unit of, 86
self inductances,
  parallel combination of, 89—90
  series combination of, 88—9
    mutually coupled, 96—7, 217—18
self inductor, 85—6
  ac behaviour of, 132—4, 139—40
  calculable, 264, 268
  complex impedance of, 139
  continuously variable, 96
  energy storage by, 87—8, 95, 99,
    104—5, 115, 133—4, 183, 265
  impedance of, 133, 140
  phase angle for, 133, 140
  power factor for, 134
  standard, 264, 279
sensitivity of bridge, 61—2, 217, 219,
  229, 240—1, 247—8, 251, 254—5
series resonance, 174—80, 201—2
shunt, universal, 50—1
siemens, the, 24
silver ampere, 261

SI system of units, 258, 263
skin effect, 164, 168, 196, 279
slab line, 163
Smith, F. E., Lorenz disc, 269—70
source, for ac bridge, 240—1
  constant—current, 27—8
  constant—voltage, 27
specific conductance, 26
specific resistance, 25, 268
speed of light, 169, 260, 262, 263, 264
standard capacitor, 82, 97, 264, 278
standard cell, 4, 43—4, 52, 54, 55—7, 264, 276, 277, 279
standard inductor, 97, 264, 279
standard mutual inductor, 97, 264, 279
standard resistor, 43, 44—5, 264, 276
star-delta, star-mesh transformation, 222, 251—6
Superposition Theorem, 243—5, 250—1
susceptance,
  definition of, 144
  measurement of, 181—2, 239
Swan, Sir Joseph, filament lamp, 7
Système International d'Unités, 258, 263

tangent galvanometer, 5
telegraphy, 6
  equations of, 165—6
temperature coefficient of resistance, 45, 49—50, 276
thermal emf, 45, 54, 61, 63, 276, 280
thermocouple instrument, 129, 280
Thévenin's Theorem, 245—8
Thompson, A. M., calculable capacitor, 272—3
Thomson, Sir William, *see* Lord Kelvin
time constant,
  of CR combination, 76—7, 79—82, 83, 116
  definition of, 76—7
  influence of resistance, 92—3
  of LCR combination, 102
  of LR combination, 91, 92, 98
torsion balance, 3
transfer admittance, 235
transfer impedance, 235, 249—50
transfer instrument, 128
transformer, 194—211 (*see also* mutual inductance)
  bridge, 231—3, 273—5, 279
  coupling factor of, 195—6, 209—11, 218
  dust-cored, 196
  effect, 6—7, 83—5
  efficiency of, 195, 196
  equivalent primary circuit of, 200—1

historic, 6—7, 83, 85
  input impedance of, 198—201, 204
  iron-cored, 194, 195, 196
  leakage inductance in, 200, 232
  loss mechanisms in, 196, 201
  open-circuit secondary voltage of, 196—7
  screened, 225, 228—30, 232
  with series-connected coils, 96—7, 217—18
  sign convention for, 93—4, 197
  tuned, 201—11
  voltage ratio of, 195
transient response,
  of CR combination, 75—82, 115—17
  of LCR combination, 99—113
    critically-damped, 106—13
    overdamped, 101—3, 109, 112—13
    underdamped, 103—6, 109, 112—13
  of LR combination, 90—3, 97—9
transmission line, 162—72
  characteristic impedance of, 164—5, 167, 168—9, 172
  coaxial, 163, 168—9
  constants of, 168—9
  dispersion in, 168
  input impedance of, 171—2
  matched, 165
  pass band of, 165
  phase-velocity in, 115, 166—8, 169
  wavelength in, 169
tungsten lamp, 7
tunnel effect, 277
Tuttle, W. N., ac bridges, 233—9, 255—6
twin-T bridge, 233—9

unbalanced transmission line, 163
undersea cable, 6
universal shunt, 50—1

vibration galvanometer, 213
volt, definition of, 8, 21, 261
Volta, historic cell, 1, 3
voltage gain, 176—7, 180, 201, 204
voltage—ratio box, 58—9
voltage resonance, 185—6, 189, 191
voltage, scale of, 13—14, 21
voltage, unit of, 8, 13—14, 20, 21, 261, 280
voltmeter, 13, 48—9, 106, 128—30

Wagner earth, 225, 227—9
wattmeter, 129
Weber electrodynamometer, 5

Weston cell, 4, 43–4, 52, 54, 55–7, 276, 277, 279
Weston moving-coil instrument, 6
Wheatstone, Sir Charles, resistance bridge, 5, 59–62, 213, 247–8, 249, 251, 254–5, 279

Wheatstone-type ac bridge, 213–31, 249
Wien bridge, 219, 221, 222

Y-delta transformation, 251–6

Zener diode, 276–7